Earth Observation Using Python

Special Publications 75

EARTH OBSERVATION USING PYTHON

A Practical Programming Guide

Rebekah B. Esmaili

This Work is a co-publication of
the American Geophysical Union and John Wiley and Sons, Inc.

AGU
ADVANCING EARTH
AND SPACE SCIENCE

WILEY

Published under the aegis of the AGU Publications Committee

Brooks Hanson, Executive Vice President, Science
Carol Frost, Chair, Publications Committee
For details about the American Geophysical Union visit us at www.agu.org.

The right of Rebekah B. Esmaili to be identified as the author of this work has been asserted in accordance with law.

Registered Office
John Wiley & Sons, Inc., 111 River Street, Hoboken, NJ 07030, USA

Editorial Office
111 River Street, Hoboken, NJ 07030, USA

For details of our global editorial offices, customer services, and more information about Wiley products visit us at www.wiley.com.

Wiley also publishes its books in a variety of electronic formats and by print-on-demand. Some content that appears in standard print versions of this book may not be available in other formats.

Library of Congress Cataloging-in-Publication Data
Name: Esmaili, Rebekah Bradley, author.
Title: Earth observation using Python : a practical programming guide / Rebekah B. Esmaili.
Description: Hoboken, NJ : Wiley, [2021] | Includes bibliographical references and index.
Identifiers: LCCN 2021001631 (print) | LCCN 2021001632 (ebook) | ISBN 9781119606888 (hardback) | ISBN 9781119606895 (adobe pdf) | ISBN 9781119606918 (epub)
Subjects: LCSH: Earth sciences–Data processing. | Remote sensing–Data processing. | Python (Computer program language) | Information visualization. | Artificial satellites in earth sciences. | Earth sciences–Methodology.
Classification: LCC QE48.8 .E85 2021 (print) | LCC QE48.8 (ebook) | DDC 550.285/5133–dc23
LC record available at https://lccn.loc.gov/2021001631
LC ebook record available at https://lccn.loc.gov/2021001632

Cover Design: Wiley
Cover Image: © NASA

MIX
Paper from responsible sources
FSC www.fsc.org FSC® C013604

CONTENTS

FOREWORD

When I first met the author a few years ago, she was eager to become more involved in the Joint Polar Satellite System's Proving Ground. The Proving Ground by definition assesses the impact of a product in the user's environment; this intrigued Rebekah because as a product developer, she wanted to understand the user's perspective. Rebekah worked with the National Weather Service to demonstrate how satellite-derived atmospheric temperature and water vapor soundings can be used to describe the atmosphere's instability to support severe weather warnings. Rebekah spent considerable time with users at the Storm Prediction Center in Norman, Oklahoma, to understand their needs, and she found their thirst for data and the need for data to be easily visualized and understandable. This is where Rebekah leveraged her expert skills in Python to provide NWS with the information they found to be most useful. Little did I know at the time she was writing a book.

As noted in this book, a myriad of Earth-observing satellites collect critical information of the Earth's complex and ever-changing environment and landscape. However, today, unfortunately, all that information is not effectively being used for various reasons: issues with data access, different data formats, and the need for better tools for data fusion and visualization. If we were able to solve these problems, then suddenly there would be vast improvements in providing societies with the information needed to support decisions related to weather and climate and their impacts, including high-impact weather events, droughts, flooding, wildfires, ocean/coastal ecosystems, air quality, and more. Python is becoming the universal language to bridge these various data sources and translate them into useful information. Open and free attributes, and the data and code sharing mindset of the Python communities, make Python very appealing.

Being involved in a number of international collaborations to improve the integration of Earth observations, I can certainly emphasize the importance of working together, data sharing, and demonstrating the value of data fusion. I am very honored to write this Foreword, since this book focuses on these issues and provides an excellent guide with relevant examples for the reader to follow and relate to.

Dr. Mitch Goldberg
Chief Program Scientist
NOAA-National Environmental Satellite, Data, and Information Service
June 22, 2020

ACKNOWLEDGMENTS

This book evolved from a series of Python workshops that I developed with the help of Eviatar Bach and Kriti Bhargava from the Department of Atmospheric and Oceanic Science at the University of Maryland. I am very grateful for their assistance providing feedback for the examples in this book and for leading several of these workshops with me.

This book would not exist without their support and contributions from others, including:

The many reviewers who took the time to read versions of this book, several of whom I have never met in person. Thanks to modern communication systems, I was able to draw from their expertise. Their constructive feedback and insights not only helped to improve this quality and breadth of the book but also helped me hone my technical writing skills.

Rituparna Bose, Jenny Lunn, Layla Harden, and the rest of the team at AGU and Wiley for keeping me informed, organized, and on track throughout this process. They were truly a pleasure to work with.

Nadia Smith and Chris Barnet, and my other colleagues at Science and Technology Corp., who provided both feedback and conversations that helped shape some of the ideas and content in this book.

Catherine Thomas, Clare Flynn, Erin Lynch, and Amy Ho for their endless encouragement and support.

Tracie and Farid Esmaili, my parents, who encouraged me to aim high even if they were initially confused when their atmospheric scientist daughter became interested in "snakes."

INTRODUCTION

Python is a programming language that is rapidly growing in popularity. The number of users is large, although difficult to quantify; in fact, Python is currently the most tagged language on stackoverflow.com, a coding Q&A website with approximately 3 million questions a year. Some view this interest as hype, but there are many reasons to join the movement. Scientists are embracing Python because it is free, open source, easy to learn, and has thousands of add-on packages. Many routine tasks in the Earth sciences have already been coded and stored in off-the-shelf Python libraries. Users can download these libraries and apply them to their research rather than simply using older, more primitive functions. The widespread adoption of Python means scientists are moving toward a common programming language and set of tools that will improve code shareability and research reproducibility.

Among the wealth of remote sensing data available, satellite datasets are particularly voluminous and tend to be stored in a variety of binary formats. Some datasets conform to a "standard" structure, such as **netCDF4.** However, because of uncoordinated efforts across different agencies and countries, such standard formats bear their own inconsistencies in how data are handled and intended to be displayed. To address this, many agencies and companies have developed numerous "quick look" methods. For instance, data can be searched for and viewed online as Jpeg images, or individual files can be displayed with free, open-source software tools like Panoply (www.giss.nasa.gov/tools/panoply/) and HDFView (www.hdfgroup.org/downloads/hdfview/).

Still, scientists who wish to execute more sophisticated visualization techniques will have to learn to code. Coding knowledge is not the only limitation for users. Not all data are "analysis ready," i.e., in the proper input format for visualization tools. As such, many pre-processing steps are required to make the data usable for scientific analysis. This is particularly evident for data fusion, where two datasets with different resolutions must first be mapped to the same grid before they are compared. Many data users are not satellite scientists or professional

---

Earth Observation Using Python: A Practical Programming Guide, Special Publications 75, First Edition. Rebekah B. Esmaili.

DOI: 10.1002/9781119606925.introduction

programmers but rather members of other research and professional communities, these barriers can be too great to overcome. Even to a technical user, the nuances can be frustrating. At worst, obstacles in coding and data visualization can potentially lead to data misuse, which can tarnish the work of an entire community.

The purpose of this text is to provide an overview of the common preparatory work and visualization techniques that are applied to environmental satellite data using the Python language. This book is highly example-driven, and all the examples are available online. The exercises are primarily based on hands-on tutorial workshops that I have developed. The motivation for producing this book is to make the contents of the workshops accessible to more Earth scientists, as very few Python books currently available target the Earth science community.

This book is written to be a practical workbook and not a theoretical textbook. For example, readers will be able to interactively run prewritten code interactively alongside the text to guide them through the code examples. Exercises in each section build on one another, with incremental steps folded in. Readers with minimal coding experience can follow each "baby step" to get them up to become "spun up" quickly, while more experienced coders have the option of working with the code directly and spending more time on building a workflow as described in Section III.

The exercises and solutions provided in this book use Jupyter Notebook, a highly interactive, web-based development environment. Using Jupyter Notebook, code can be run in a single line or short blocks, and the results are generated within an interactive documented format. This allows the student to view both the Python commands and comments alongside the expected results. Jupyter Notebook can also be easily converted to programs or scripts than can be executed on Linux Machines for high-performance computing. This provides a friendly work environment to new Python users. Students are also welcome to develop code in any environment they wish, such as the Spyder IDE or using iPython.

While the material builds on concepts learned in other chapters, the book references the location of earlier discussions of the material. Within each chapter, the examples are progressive. This design allows students to build on their understanding knowledge (and learn where to find answers when they need guidance) rather than memorizing syntax or a "recipe." Professionally, I have worked with many datasets and I have found that the skills and strategies that I apply on satellite data are fairly universal. The examples in this book are intended to help readers become familiar with some of the characteristic quirks that they may encounter when analyzing various satellite datasets in their careers. In this regard, students are also strongly encouraged to submit requests for improvements in future editions.

Like many technological texts, there is a risk that the solutions presented will become outdated as new tools and techniques are developed. The sizable user community already contributing to Python implies it is actively advancing; it is a living language in contrast to compiled, more slowly evolving legacy languages like

Fortran and C/C++. A drawback of printed media is that it tends to be static and Python is evolving more rapidly than the typical production schedule of a book. To mitigate this, this book intends to teach fluency in a few, well-established packages by detailing the steps and thought processes needed for a user needs to carry out more advanced studies. The text focuses discipline-agnostic packages that are widely used, such as **NumPy, Pandas,** and **xarray,** as well as plotting packages such as **Matplotlib** and **Cartopy.**

I have chosen to highlight Python primarily because it is a general-purpose language, rather than being discipline or task-specific. Python programmers can script, process, analyze, and visualize data. Python's popularity does not diminish the usefulness and value of other languages and techniques. As with all interpreted programming languages, Python may run more slowly compared to compiled languages like Fortran and C++, the traditional tools of the trade. For instance, some steps in data analysis could be done more succinctly and with greater computational efficiency in other languages. Also, underlying packages in Python often rely on compiled languages, so an advanced Python programmer can develop very computationally efficient programs with popular packages that are built with speed-optimized algorithms. While not explicitly covered in this book, emerging packages such as Dask can be helpful to process data in parallel, so more advanced scientific programmers can learn to optimize the speed performance of their code. Python interfaces with a variety of languages, so advanced scientific programmers can compile computationally expensive processing components and run them using Python. Then, simpler parts of the code can be written in Python, which is easier to use and debug.

This book encourages readers to share their final code online with the broader community, a practice more common among software developers than scientists. However, it is also good practice to write code and software in a thoughtful and carefully documented manner so that it is usable for others. For instance, well-written code is general purpose, lacks redundancy, and is intuitively organized so that it may be revised or updated if necessary. Many scientific programmers are self-learners with a background in procedural programming, and thus their Python code will tend to resemble the flow of a Fortran or IDL program. This text uses Jupyter Notebook, which is designed to promote good programming habits in establishing a "digestible code" mindset; this approach organizes code into short chunks. This book focuses on clear documentation in science algorithms and code. This is handled through version control, using virtual environments, how to structure a usable README file, and what to include in inline commenting.

For most environmental science endeavors, data and code sharing are part of the research-to-operations feedback loop. "Operations" refers to continuous data collection for scientific research and hazard monitoring. By sharing these tools with other researchers, datasets are more fully and effectively utilized. Satellite data providers can upgrade existing datasets if there is a demand. Globally, satellite data are provided through data portals by NASA, NOAA, EUMETSAT, ESA, JAXA, and other international agencies. However, the value of these datasets is often only visible through scientific journal articles, which only represent a small subset of potential users. For instance, if the applications of satellite observations used for routine disaster mitigation and planning in a disadvantaged nation are not published in a scientific journal, improvements for disaster-mitigation specific needs may never be met.

Further, there may be unexpected or novel uses of datasets that can drive scientific inquiry, but if the code that brings those uses to life is hastily written and not easily understood, it is effectively a waste of time for colleagues to attempt to employ such applications. By sharing clearly written code and corresponding documentation for satellite data applications, users can alert colleagues in their community of the existence of scientific breakthrough efforts and expand the potential value of satellite datasets within and beyond their community. Moreover, public knowledge of those efforts can help justify the versatility and value of satellite missions and provide a return on investment for organizations that fund them. In the end, the dissemination of code and data analysis tools will only benefit the scientific community as a whole.

Part I

Overview of Satellite Datasets

1

A TOUR OF CURRENT SATELLITE MISSIONS AND PRODUCTS

There are thousands of datasets containing observations of the Earth. This chapter describes some satellite types, orbits, and missions, which benefit a variety of fields within Earth sciences, including atmospheric science, oceanography, and hydrology. Data are received on the ground through receiver stations and processed for use using retrieval algorithms. But the raw data requires further manipulation to be useful, and Python is a good choice for analysis and visualization of these datasets.

At present, there are over 13,000 satellite-based Earth observations freely and openly listed on www.data.gov. Not only is the quantity of available data notable, its quality is equally impressive; for example, infrared sounders can estimate brightness temperatures within 0.1 K from surface observations (Tobin et al., 2013), imagers can detect ocean currents with an accuracy of 1.0 km/hr (NOAA, 2020), and satellite-based lidar can measure the ice-sheet elevation change with a 10 cm sensitivity (Garner, 2015). Previously remote parts of our planet are now observable, including the open oceans and sparsely populated areas. Furthermore, many datasets are available in near real time with image latencies ranging from less than an hour down to minutes – the latter being critically important for natural disaster prediction. Having data rapidly available enables science applications and weather prediction as well as to emergency management and disaster relief. Research-grade data take longer to process (hours to months) but has a higher accuracy and precision, making it suitable for long-term consistency. Thus, we live in the "golden age" of satellite Earth observation. While the data are accessible, the tools and skills necessary to display and analyze this information require practice and training.

Earth Observation Using Python: A Practical Programming Guide, Special Publications 75, First Edition. Rebekah B. Esmaili.

DOI: 10.1002/9781119606925.ch1

Python is a modern programming language that has exploded in popularity, both within and beyond the Earth science community. Part of its appeal is its easy-to-learn syntax and the thousands of available libraries that can be synthesized with the core Python package to do nearly any computing task imaginable. Python is useful for reading Earth-observing satellite datasets, which can be difficult to use due to the volume of information that results from the multitude of sensors, platforms, and spatio-temporal spacing. Python facilitates reading a variety of self-describing binary datasets in which these observations are often encoded. Using the same software, one can complete the entirety of a research project and produce plots. Within a notebook environment, a scientist can document and distribute the code to other users, which can improve efficiency and transparency within the Earth sciences community.

Satellite data often require some pre-processing to make it usable, but which steps to take and why are not always clear. Data users often misinterpret concepts such as data quality, how to perform an atmospheric correction, or how to implement the complex gridding schemes necessary to compare data at different resolutions. Even to a technical user, the nuances can be frustrating and difficult to overcome. This book walks you through some of the considerations a user should make when working with satellite data.

The primary goal of this text is to get the reader up to speed on the Python coding techniques needed to perform research and analysis using satellite datasets. This is done by adopting an example-driven approach. It is light on theory but will briefly cover relevant background in a nontechnical manner. Rather than getting lost in the weeds, this book purposefully uses realistic examples to explain concepts. I encourage you to run the interactive code alongside reading the text. In this chapter, I will discuss a few of the satellites, sensors, and datasets covered in this book and explain why Python is a great tool for visualizing the data.

1.1. History of Computational Scientific Visualization

Scientific data visualizing used to be a very tedious process. Prior to the 1970s, data points were plotted by hand using devices such as slide rules, French curls, and graph paper. During the 1970s, IBM mainframes became increasingly available at universities and facilitated data analysis on the computer. For analysis, IBM mainframes required that a researcher write Fortran-IV code, which was then printed to cards using a keypunch machine (Figure 1.1). The punch cards then were manually fed into a shared university computer to perform calculations. Each card is roughly one line of code. To make plots, the researcher could create a Fortran program to make an ASCII plot, which creates a plot by combining lines, text, and symbols. The plot could then be printed to a line-printer or a teleprinter. Some institutions had computerized graphic devices, such as Calcomp plotters. Rather than create ASCII plots, the researcher could use a Calcomp plotting

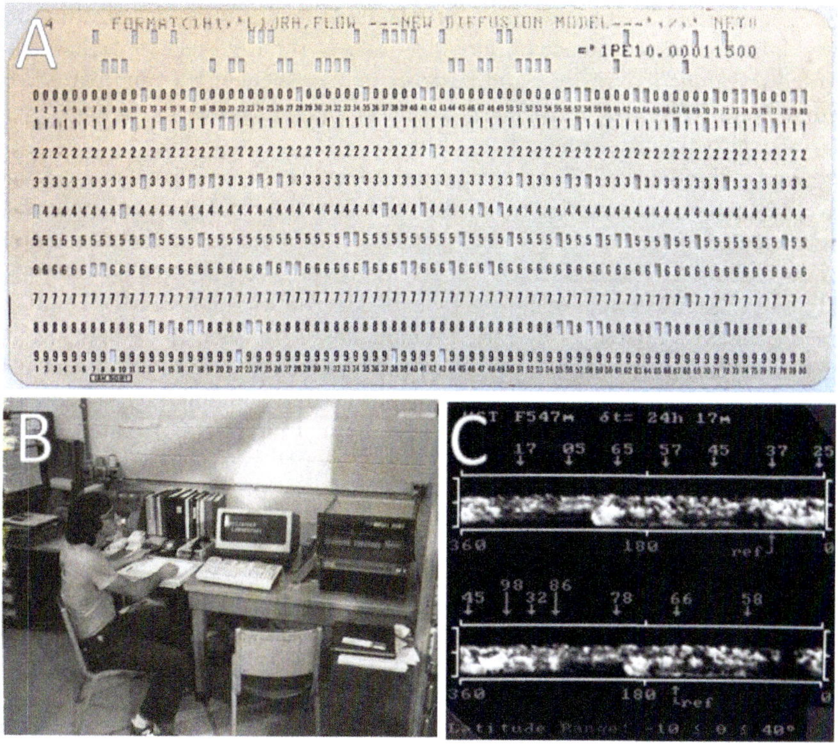

Figure 1.1 (a) An example of a Fortran punch card. Each vertical column represents a character and one card roughly one line of Fortran code. (b) 1979 photo of an IMSAI 8080 computer that could store up to 32 kB of the data, which could then be transferred to a keypunch machine to create punch cards. (c) an image created from the Hubble Space Telescope using a Calcomp printer, which was made from running punch cards and plotting commands through a card reader.

command library to control how data were visualized and store the code on computer tape. The scientist would then take the tape to a plotter, which was not necessarily (or usually) in the same area as the computer or keypunch machine. Any errors – such as bugs in the code, damaged punch cards, or damaged tape – meant the whole process would have to be repeated from scratch.

In the mid-1980s, universities provided remote terminals that would eventually replace the keypunch and card reader machine system. This substantially improved data visualization processes, as scientists no longer had to share limited resources such as keypunch machines, card readers, or terminals. By the late 1980s, personal computers became more affordable for scientists. A typical PC, such as the IBM XT 286, had 640 Kb of random access memory, a 32 MB hard drive, and 5.25 inch floppy disks with 1.2 MB of disk storage (IBM, 1989). At this time, pen plotters became increasingly common for scientific visualization, followed later by the prevalence of ink-jet printers in the 1990s. These technologies allowed researchers to process and visualize data conveniently from their offices. With the proliferation of user-friendly person computers, printers eventually made their way into all homes and offices.

Now with advances in computing and internet access, researchers no longer need to print their visualizations at all, but often keep data in digital form only. Plots can be created in various data formats that easily embed into digital presentations and documents. Scientists often do not ever print visualizations because computers and cloud storage can store many gigabytes of data. Information is created and consumed entirely in digital form. Programming languages, such as Python, can tap into high-level plotting programs and can minimize the axis calculation and labeling requirements within a plot. Thus, the expanded access to computing tools and simplified processes have advanced scientific data visualization opportunities.

1.2. Brief Catalog of Current Satellite Products

In Figure 1.2, you can see that the international community has developed and launched a plethora of Earth-observing satellites, each with several onboard sensors that have a range of capabilities. I am not able to discuss every sensor, dataset, and **mission** (a term coined by NASA to describe projects involving

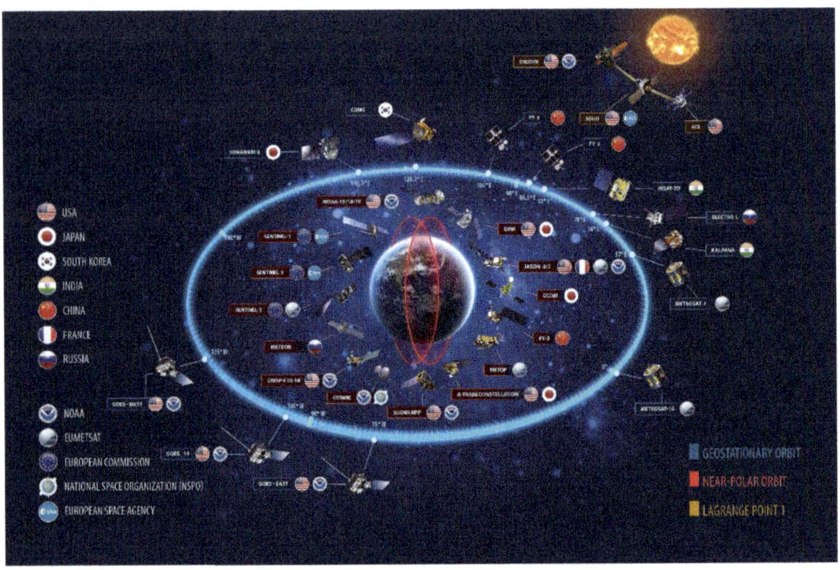

Figure 1.2 Illustration of current Earth, space weather, and environmental monitoring satellites from the World Meteorological Organization (WMO). Source: U.S. Department of Commerce / NOAA / Public Domain.

spacecraft). However, I will describe some that are relevant to this text, organized by subject area.

1.2.1. Meteorological and Atmospheric Science

Most Earth-observing satellites orbit our planet either in either **geostationary** or **low-Earth orbiting** patterns. These types of satellites tend to be managed and operated by large international government agencies, and the data are often freely accessible online:

- **Geosynchronous equatorial orbit (GEO) satellites.** Geostationary platforms orbit the Earth at 35,700 km above the Earth's surface. GEO satellites are designed to continuously monitor the same region on Earth, and thus can provide many images over a short period of time to monitor change. National Oceanic and Atmospheric Administration (NOAA) operates the Geostationary Environmental Satellite System (GOES) satellites for monitoring North and South America. GOES-16 and -17 have an **advanced baseline instrument** (ABI) onboard to create high-resolution imagery in **visible** and **infrared** (IR) wavelengths. The GOES-16 and -17 satellites are also equipped with the Geostationary Lightning Mapper (GLM) to detect lightning. Instruments designed for space weather include the Solar Ultraviolet Imager (SUVI) and X-ray Irradiance Sensors (EXIS). The European Organization for the Exploitation of Meteorological Satellites (EUMETSAT) operates and maintains the **Meteosat** series GEO satellites that monitor Europe and Africa. The Japan Aerospace Exploration Agency (JAXA) operates and maintains the **Himawari** satellite that monitors Asia and Oceania.
- **Low-Earth orbit (LEO) satellites.** Polar orbiting satellites provide approximately twice daily global observations at the equator (with more observations per day at the poles). Figure 1.3 displays the **equatorial crossing time** for historic and existing LEO satellites, which refers to the local time at the equator when observations are made. Overpasses from some LEO satellites shift during a mission, while others are periodically adjusted back to maintain a consistent overpass time throughout the duration of a mission. Polar orbiting satellites are called low-Earth orbit satellites because they are much closer to the Earth's surface (at 400–900 km) than GEO satellites, which are approximately 40 times further away from the earth or at ~35,000 km. The lower altitude of LEO satellites facilitates their higher spatial resolution relative to GEO, although the temporal resolution tends to be lower for LEO satellites. The **Suomi-NPP** and **NOAA-20** are two satellites that were developed and maintained by NASA and NOAA, respectively. They are each equipped with an **imager**, the Visible Infrared Imaging Radiometer Suite (VIIRS), and infrared and microwave **sounders**, the Cross-track Infrared Sounder (CrIS) and an Advanced Technology Microwave Sounder (ATMS). The **MetOp** series of LEO satellites (named MetOp-A, -B, and -C) were developed by the European Space Agency (ESA) and are operated by EUMETSAT.

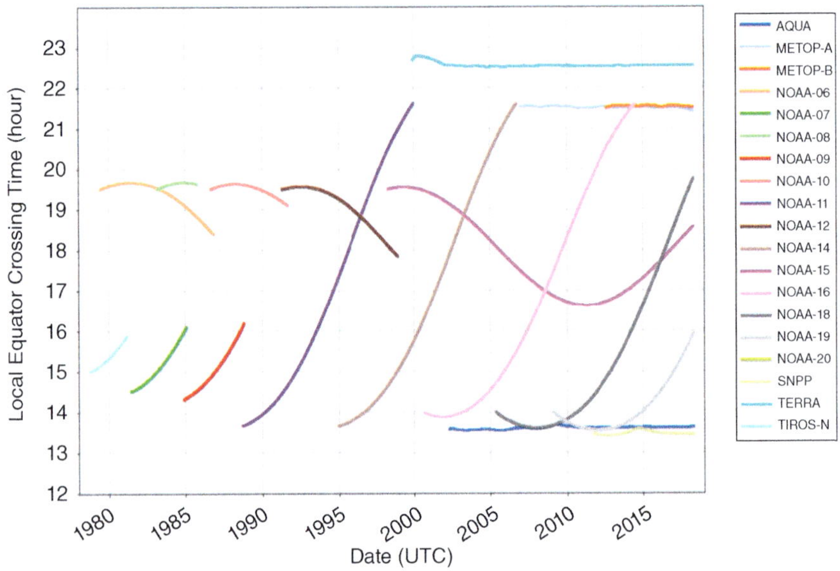

Figure 1.3 Equatorial crossing times for various LEO satellites displayed using Python.

1.2.2. Hydrology

Because water is sensitive to microwave frequencies, microwave instruments and sounders are useful for detecting water vapor, precipitation, and ground moisture. The Global Precipitation Mission (GPM) uses the core GPM satellite along with a constellation of microwave imagers and sounders to estimate global precipitation. The SMAP satellite mission uses active and passive microwave sensors to observe surface soil moisture every two to three days. The GRACE-FO satellite measures gravitational anomalies, that can be used to infer changes in global sea levels and soil moisture. All three hydrology missions were developed and operated by NASA.

1.2.3. Oceanography and Biogeosciences

Both GEO and LEO satellites can provide sea surface temperature (SST) observations. The GOES series of GEO satellites provides continuous sampling of SSTs over the Atlantic and Pacific Ocean basins. The MODIS instrument on the Aqua satellite has been providing daily, global SST observations continuously since the year 2000. Visible wavelengths are useful for detecting ocean color, particularly from LEO satellites, which are often observed at very high resolutions.

Additionally, LEO satellites can detect global sea-surface anomaly parameters. **Jason-3** is a low-Earth satellite developed as a partnership between EUMETSAT, NOAA, NASA, and CNES. The radar altimeter instrument on Jason-3 is sensitive to height changes less than 4 cm and completes a full Earth scan every 10 days (Vaze et al., 2010).

1.2.4. Cryosphere

ICESat-2 (Ice, Cloud, and land Elevation Satellite 2) is a LEO satellite mission designed to measure ice sheet elevation and sea ice thickness. The GRACE-FO satellite mission can also monitor changes in glaciers and ice sheets.

1.3. The Flow of Data from Satellites to Computer

The missions mentioned in the previous section provide open and free data to all users. However, **data delivery**, the process of downloading sensor data from the satellite and converting it into a usable form, is not trivial. Raw sensor data are first acquired on the satellite, then the data must be relayed to the Earth's ground system, often at speeds around 30 Mbits/second. For example, GOES satellite data are acquired by NASA's Wallops Flight Facility in Virginia; data from the Suomi NPP satellite is downloaded to the ground receiving station in Svalbard, Norway (Figure 1.4). Once downloaded, the observations are calibrated and several corrections are applied, such as an atmospheric correction to reduce haze in the image or topographical corrections to adjust changes in pixel brightness on complex terrain. The corrected data are then incorporated into physical products using satellite **retrieval algorithms.** Altogether, the speed of data download and processing can impact the **data latency**, or the difference between the time the physical observation is made and the time it becomes available to the data user.

Data can be accessed in several ways. The timeliest data can be downloaded using a direct broadcast (DB) antenna, which can immediately receive data when the satellite is in range. This equipment is expensive to purchase and maintain, so usually only weather and hazard forecasting offices install them. Most users will access data via the internet. FTP websites post data in near real time, providing the data within a few hours of the observation. Not all data must be timely – research-grade data can take months to calibrate to ensure accuracy. In this case, ordering through an online data portal will grant users access to long records of data.

While data can be easily accessed online, they are rarely analysis ready. Software and web-based tools allow for quick visualization, but to create custom analyses and visualizations, coding is necessary. To combine multiple datasets, each must be gridded to the same resolution for an apples-to-apples comparison. Further, data providers use quality flags to indicate the likelihood of a suitable retrieval. However, the meaning and appropriateness of these flags are not always

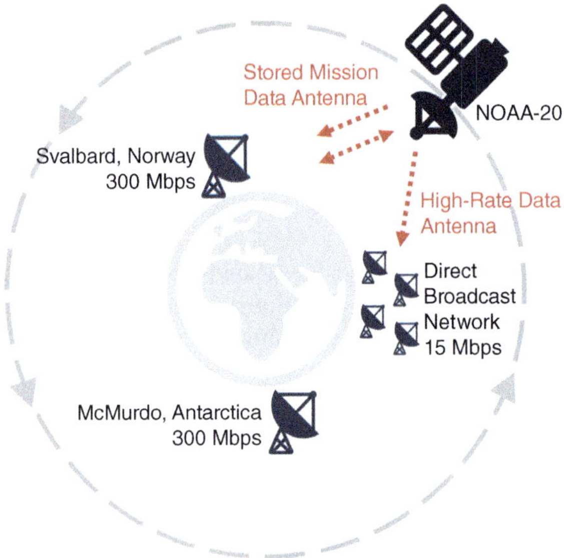

Figure 1.4 NOAA-20 satellite downlink.

well communicated to data users. Moreover, understanding how such datasets are organized can be cumbersome to new users. This text thus aims to identify specific Python routines that enable custom preparation, analysis, and visualization of satellite datasets.

1.4. Learning Using Real Data and Case Studies

I have structured this book so that you can learn Python through a series of examples featuring real phenomena and public datasets. Some of the datasets and visualizations are useful for studying wildfires and smoke, dust plumes, and hurricanes. I will not cover all scenarios encountered in Earth science, but the skills you learn should be transferrable to your field. Some of these case studies include:

- **California Camp Fire (2018).** California Camp Fire was a forest fire that began on November 8, 2018, and burned for 17 days over a 621 km^2 area. It was primarily caused by very low regional humidity due to strong gusting wind events and a very dry surface. The smoke from the fire also affected regional air quality. In this case study, I will examine satellite observations to show the location and intensity as well as the impact that the smoke had on regional CO, ozone, and aerosol optical depth (AOD). Combined satellite channels also provide useful imagery for tracking smoke, such as the dust

RGB product. Land datasets such as the Normalized Difference Vegetation Index (NDVI) are useful for highlighting burn scars from before and after the fire events.

- **Hurricane Michael (2018).** Michael was a major hurricane that affect the Florida Panhandle of the United States. Michael developed as a tropical wave on October 7 in the southwest Caribbean Sea and grew into a Category 5 storm by October 10. Throughout its life cycle, Michael caused extensive flooding, leading to 74 deaths and $25 billion in damage. Several examples in this text utilize visible and infrared imagery of Hurricane Michael.
- **Louisiana Flood Event (2016).** Thousands of homes were flooded in Louisiana when over 20 inches of rain fell between August 12 and August 21, 2016. The event began after a mesoscale convective system stalled over the area near Baton Rouge and Lafayette, Louisiana. I will use the IMERG global rainfall dataset to examine this event.

1.5. Summary

I have provided a brief overview of the many satellite missions and datasets that are available. This book has two main objectives: (1) to make satellite data and analysis accessible to the Earth science community through practical Python examples using real-world datasets; and (2) to promote a reproducible and transparent scientific code philosophy. In the following chapters, I will focus on describing data conventions, common methods, and problem-solving skills required to work with satellite datasets.

References

Tobin, D., Revercomb, H., Knuteson, R., Taylor, J., Best, F., Borg, L., et al. (2013). Suomi-NPP CrIS radiometric calibration uncertainty. *Journal of Geophysical Research: Atmospheres, 118*(18), 10,589–10,600. https://doi.org/10.1002/jgrd.50809

Garner, R. (2015, July 10). ICESat-2 Technical Requirements. www.nasa.gov/content/goddard/icesat-2-technical-requirements

IBM (1989, January). Personal Computer Family Service Information Manual. IBM document SA38-0037-00. http://bitsavers.trailing-edge.com/pdf/ibm/pc/SA38-0037-00_Personal_Computer_Family_Service_Information_Manual_Jul89.pdf

National Oceanic and Atmospheric Administration (2020, June 12). GOES-R Series Level I Requirements (LIRD). www.goes-r.gov/syseng/docs/LIRD.pdf.

2

OVERVIEW OF PYTHON

Python is a free and open-source programming language. There are over 200,000 packages registered online that expand Python's capabilities. This chapter provides a description of some useful packages for the Earth sciences. Some of these useful packages include NumPy, Pandas, Matplotlib, netCDF4, h5py, Cartopy, and xarray. These packages have a strong development base and a large community of support, making them appropriate for scientific investigation.

In this chapter, I discuss some reasons why Python is a valuable tool for Earth scientists. I will also provide an overview of some of the commonly used Python packages for remote sensing applications that I will use later in this book. Python evolves rapidly, so I expect these tools to improve and new ones to become available. However, these will provide a solid foundation for you to begin your learning.

2.1. Why Python?

Chances are, you may already know a little about what Python is and have some motivation to learn it. Below, I outline common reasons to use Python relevant to the Earth sciences:

- **Python is open-source and free.** Some of the legacy languages are for profit, and licenses can be prohibitively expensive for individuals. If your career plans include remaining at your current institution or company that supplies you with language licenses, then open source may not be of concern to you. But often, career growth means working for different organizations. Python is

Earth Observation Using Python: A Practical Programming Guide, Special Publications 75, First Edition. Rebekah B. Esmaili.

DOI: 10.1002/9781119606925.ch2

portable, which frees up your skillset from being exclusively reliant on proprietary software.

- **Python can increase productivity.** There are thousands of supported libraries to download and install. For instance, if you want to open multiple netCDF files at once, the package called **xarray** can do that. If you want to re-grid an irregular dataset, there is a package called **pyresample** that will do this quickly. Even more subject-specific plots, like Skew-T diagrams, have a prebuilt package called **MetPy.** For some datasets, you can download data directly into Python using **OPenDAP.** Overall, you spend less time developing routines and more time analyzing results.
- **Python is easy to learn, upgrade, and share.** Python code is very "readable" and easy to modularize, so that functions can be easily expanded or improved. Further, low- or no-cost languages like Python increase the shareability of the code. When the code works, it should be distributed online for other users' benefit. In fact, some grants and journals require online dissemination.

You may already have knowledge of other computer languages such as IDL, MATLAB, Fortran, C++, or R. Learning Python does not mean you will stop using other languages or rewrite all your existing code. In many cases, other languages will continue to be a valuable part of your daily work. For example, there are a few drawbacks to using Python:

- **Python may be slower than compiled languages.** While many of the core scientific packages compile code on the back end, Python itself is not a compiled language. For a novice user, Python will run more slowly, especially if loops are present in the code. For a typical user, this speed penalty may not be noticeable, and advanced users can tap into other runtime frameworks like Dask or Cython or even run compiled Fortran subroutines to enhance performance. However, new users might not feel comfortable learning these workarounds, and even with runtime frameworks and subroutines, performance might not improve. If speed is a concern, then Python could be used as a prototype code tool prior to converting into a compiled language.
- **New users often run packages "as-is" and the contents are not inspected.** There are thousands of libraries available, but many are open-source, community projects and bugs and errors will exist. For example, irregular syntax can result whenever there is a large community of developers. Thus, scientists and researchers should be extra vigilant and only use vetted packages.
- **Python packages may change function syntax or discontinued.** Python changes rapidly. While most developers refrain from abruptly changing syntax, this practice is not always followed. In contrast, because much of the work in developing these packages is on a volunteer basis, the communities supporting them could move on to other projects and those who take over could begin a completely new syntax structure. While this is unlikely to be the case for highly

used packages, anything is possible. For example, a popular map plotting package called **Basemap** was discontinued in the module **Matplotlib** and replaced with **Cartopy**, which is an older code package. I recommend using packages that have backing from Earth science research institutions (e.g., the UK Met Office, NASA, Lamont, etc.) to raise confidence that the packages you choose to use will be relatively stable.

Unlike legacy languages such as Fortran and C++, there is no guarantee that code written in Python will remain stable for 30+ years. However, the packages presented in this book are "mature" and are likely to continue to be supported for many years. Additionally, you can reproduce the exact packages and versions using virtual environments (Section 11.3). This text highlights newer packages that save significant amounts of development time and streamline certain processes, including how to open and read **netCDF** files and gridding operations.

2.2. Useful Packages for Remote Sensing Visualization

Python contains intrinsic structure and mathematical commands, but its capabilities can be dramatically increased using **modules**. Modules are written in Python or a compiled language like C to help simplify common, general, or redundant tasks. For instance, the datetime module helps programmers manipulate calendar dates and times using a variety of units. **Packages** contain one or more modules, which are often designed to facilitate tasks that follow a central theme. Some other terms used interchangeably for packages are libraries and distributions.

At the time of writing, there are over 200,000 Python packages registered on pipy.org and more that live on the internet in code repositories such as GitHub (https://github.com/). Many of the most popular packages are often developed and maintained by large online communities. This massive effort benefits you as a scientist because many common tasks have already been developed in Python by someone else. This can also create a dilemma for scientists and researchers – the trade-off between using existing code to save time against time spent researching and vetting so many code options. Additionally, because many of these packages do not have full-time staff support, the projects can be abandoned by their development teams, and your code could eventually become obsolete.

In your research, I suggest you use three rules when choosing packages to learn and work with:

1. Use established packages.
2. Use packages that have a large community of support.
3. Use code that is efficient with respect to reduced coding time and increased speed of performance.

Following is a list of the main Python packages that I will cover in this text.

2.2.1. NumPy

NumPy is the fundamental package for scientific computing with Python. It can work with multidimensional arrays, contains many advanced mathematical functions, and is useful for linear algebra, Fourier transforms, and for generating random numbers. NumPy also allows users to encapsulate data efficiently. If you are familiar with MATLAB, you will feel very comfortable using this package.

2.2.2. Pandas

Pandas is a library that permits using data frame (stylized DataFrame) structures and includes a suite of I/O and data manipulation tools. Unlike NumPy, Pandas allows you to reference named columns instead of using indices. With Pandas, you can perform the same kinds of essential tasks that are available in spreadsheet programs (but now automated and with fewer mouse clicks!). For those who are familiar with R programming language, Pandas mimics the R data.frame function.

A limitation of Pandas is that it can only operate with 2D data structures. More recently, the **xarray** package has been developed to handle higher-dimensional datasets. In addition, Pandas can be somewhat inefficient because the library is technically a wrapper for NumPy, so it can consume up to three times as much memory, particularly in Jupyter Notebook. For larger row operations (500K rows or greater), the differences can even out. (Goutham, 2017).

2.2.3. Matplotlib

Matplotlib is a plotting library, arguably the most popular one. Matplotlib can generate histograms, power spectra, bar charts, error charts, and scatterplots with a few lines of code. The plots can be completely customized to suit your aesthetics. Due to their similarities, this is another package where MATLAB experience may come in handy.

2.2.4. netCDF4 and h5py

I will discuss two common self-describing data formats, netCDF and HDF, in Section 3.2.3. Two major packages for importing these formats are the **netCDF4** and **h5py** packages. These tools are advantageous because the user does not have to have any knowledge of how to parse the input files, so long as the files follow standard formatting. These two packages import the data, which can then be converted to NumPy to perform more rigorous data operations.

2.2.5. Cartopy

Cartopy is a package for projecting geospatial data to maps. It can also be used to access a wealth of features, including land/ocean masks and topography. Many projections are available, and you can easily transform the data between them.

Previously, **Basemap** was the primary package for creating maps. You may come across examples that use it online. However, the package is now deprecated and Cartopy has become the primary package that interfaces with Matplotlib.

Cartopy is a package available from the SciTools organization, which was originally developed by the UK Met Office. It has now expanded into a community collaboration.

2.3. Maturing Packages

The packages detailed in this section are worth mentioning because they may apply to your specific project. Further, some features are too good to ignore, so they are highlighted below. However, if your code requires a long-term shelf life, it may be best to find alternative solutions, as the following packages may change more rapidly than those listed in Section 2.3.

2.3.1. xarray

xarray is a package that borrows heavily from Pandas to organize multidimensional data. Mathematical operations are lightning fast thanks to dimensional and coordinate indexing. Visualization is also easy. xarray is valuable to Earth scientists because it permits opening multiple netCDF files with ease. Interpolation and group operations is also possible.

The xarray syntax can be challenging to newcomers. It can be difficult to wrangle the data into the format needed. Nevertheless, this tool is worth the time investment due to the many features of interest to Earth science.

2.3.2. Dask

Dask interfaces with Pandas, Scikit-Learn, and NumPy to perform parallel processing and out-of-memory operations that can read data in chunks without ever being totally in the computer's RAM. This is very useful for working with large datasets. If speed needs to be prioritized, it would be worth learning this package.

2.3.3. Iris

Iris is a format-agnostic Python library for analyzing and visualizing Earth science data. If datasets follow the standard CF formatting conventions, Iris can easily load the data. The Iris package has a steep learning curve but can be useful for performing meteorological computations. Like Cartopy, Iris is a package available from the SciTools organization.

2.3.4. MetPy

MetPy is a collection of tools in Python for reading, visualizing, and performing calculations with weather data. MetPy enables downloading a curated collection of remote sensing datasets. Unit conversions are easy to perform, which is helpful when making calculations of meteorological variables. MetPy is maintained by Unidata in Boulder, Colorado.

2.3.5. cfgrib and eccodes

Cfgrib is a useful package for reading GRIB1 and GRIB2 data, which is a common format for reanalysis and model data, particularly for the ECMWF. Cfgrib decodes GRIB data in a way that it mimics the structure of NetCDF files using the **ecCodes** python package. ecCodes was developed by ECMWF to decoding and encoding standard WMO GRIB and BUFR files.

2.4. Summary

I hope you are excited to begin your Python journey. Since it is free and open-source, Python is a valuable tool that you can carry with you for the rest of your career. Furthermore, there are many existing packages to perform common tasks in the Earth Sciences, such as importing common datasets, organizing data, performing mathematical analysis, and displaying results. In the next chapter, I will describe some of the common satellite data formats you may encounter.

References

Dask: Scalable analytics in Python. (n.d.). Retrieved November 25, 2020, from https://dask.org/

ecmwf/cfgrib. (2020). Python, European Centre for Medium-Range Weather Forecasts. Retrieved from https://github.com/ecmwf/cfgrib (Original work published July 16, 2018).

Matplotlib: Python plotting — Matplotlib 3.3.3 documentation. (n.d.). Retrieved November 25, 2020, from https://matplotlib.org/.

MetPy — MetPy 0.12. (n.d.). Retrieved November 25, 2020, from https://unidata.github.io/MetPy/latest/index.html.

NumPy. (n.d.). Retrieved November 25, 2020, from https://numpy.org/

Overview: Why xarray? — xarray 0.16.2.dev3+g18a59a6d.d20200920 documentation. (n.d.). Retrieved November 25, 2020, from http://xarray.pydata.org/en/stable/why-xarray.html.

pandas documentation — pandas 1.1.4 documentation. (n.d.). Retrieved November 25, 2020, from https://pandas.pydata.org/pandas-docs/stable/index.html.

Vaze, P., Neeck, S., Bannoura, W., Green, J., Wade, A., Mignogno, M., et al. (2010). The Jason-3 Mission: completing the transition of ocean altimetry from research to operations. In R. Meynart, S. P. Neeck, & H. Shimoda (Eds.) (p. 78260Y). Presented at the Remote Sensing, Toulouse, France. https://doi.org/10.1117/12.868543.

3

A DEEP DIVE INTO SCIENTIFIC DATA SETS

Satellite data is voluminous, so data must be compressed for storage while also documenting the contents. Scientific data can be stored in text formats that are human readable but lack a high degree of compression. Binary data are highly compressed but not human readable. Furthermore, unless supporting documentation is included in the data archive, it may be impossible to know how to read the files or what kind of data are stored in them. Self-describing formats like NetCDF4 and HDF5 contain compressed data but also store metadata inside the data file, which helps users understand the contents, such as the description of the data and data quality.

NASA's Earth Observing System Data and Information System (EOSDIS) has accumulated 27 PBs of data since the 1990s with the purpose of furthering scientific research. EOSDIS continues to add data from missions prior to the 1990s, which are stored as hard disk media (Figure 3.1). Many of these older datasets need to be "rescued," which is challenging because such legacy media are often disorganized, lack documentation, are physically damaged, or lack appropriate readers (Meier et al., 2013; James et al., 2018). As the satellite era began in the 1960s, it is unlikely that the planners considered how voluminous the data would become and how rapidly it would be produced.

Nowadays, data archiving at agencies is carefully planned and organized, and it uses **scientific data formats**. This infrastructure allows EOSDIS to freely distribute hundreds of millions of data files a year to the public (I describe how to obtain data in Appendix E). In addition to improving access and storage, scientific data formats provide consistency between files, which reduces the burden on researchers who can more easily write code using tools like Python to read, combine, and analyze data from different sources. This philosophy is encompassed by the term

---

Earth Observation Using Python: A Practical Programming Guide, Special Publications 75, First Edition. Rebekah B. Esmaili.

DOI: 10.1002/9781119606925.ch3

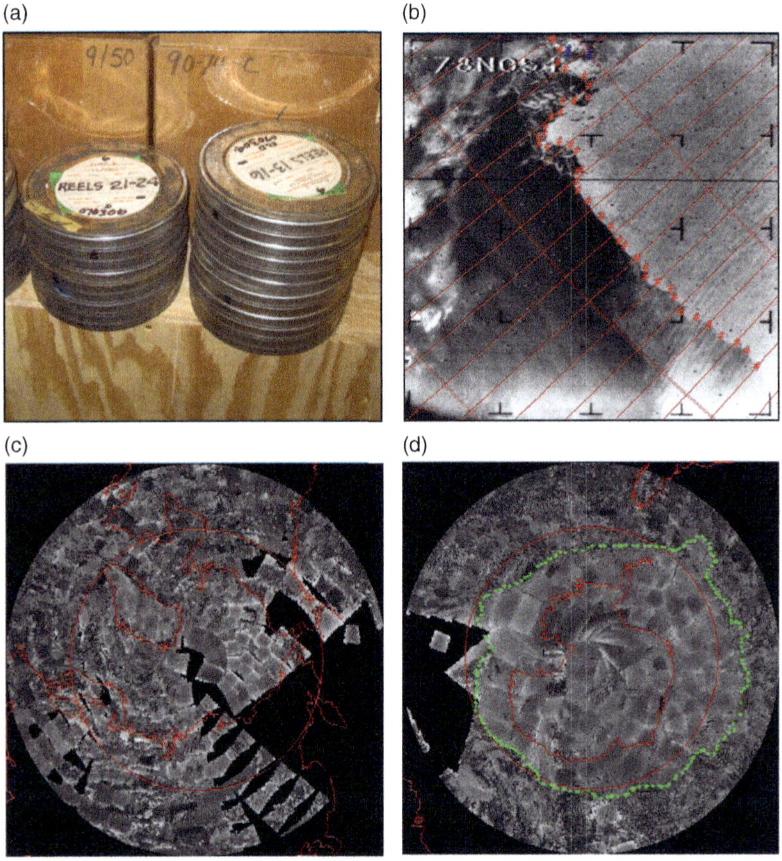

Figure 3.1 (a) Canisters of 35mm film that contain imagery recovered from Nimbus 1 of polar sea ice extent collected in 1964. From roughly 40,000 recovered images, scientists were able to reconstruct scenes such as (b) the ice edge north of Russia (78°N, 54°E) and composites of the (c) north and (d) south poles. Photos courtesy of the National Snow and Ice Data Center, University of Colorado, Boulder.

analysis-ready data (Dwyer et al., 2018). Common formats and labeling also ensure that data are well understood and used appropriately. In this chapter, I go over some of the ways data are stored, provide an overview of major satellite data formats, as well as common ways to classify and label satellite data.

3.1. Storage

3.1.1. Single Values

Satellite data are voluminous, so data need to be stored as computationally efficiently as possible. While detailed discussion is beyond the scope of this text, computers store data in bits for short, and there are 8 bits in a byte. Table 3.1

Table 3.1 Data Types, Typical Ranges, and Decimal Precision and Size in Computer Memory

Numeric Type	Range of Values	Decimal Precision
1-byte integer	–128 to 127 (0–255)	–
2-byte integer	–32,768 to 32,767 (0– to 65,535)	–
4-byte integer	–2,147,483,648 to 2,147,483,647 (0–4,294,967,295)	–
4-byte float	-3.4×10^{-38} to 3.4×10^{38}	6
8-byte float	-1.7×10^{-308} to 1.7×10^{308}	15

Note: Numbers in parentheses are the unsigned integers (positive only). Unsigned floats are not supported in Python.

has some computer number formats that satellite data are commonly stored in, along with their respective numeric ranges and storage requirements. **Integers** are numbers that have no decimal precision (e.g., 4, 8, 15, 16, –5,250, 8,642, …). **Floats** are real numbers that have decimal precision (e.g., 3.14, 5.0, 1.2E23). Integers are typically advantageous for storage because they are smaller and will not have rounding errors like float values can. To keep data small, values are stored in the smallest numerical type necessary.

Even if an observed value is large, values can be linearly scaled (using an **offset** and a **scale factor**) to fit within integer ranges to keep file sizes small. This is because many Earth observations naturally only scale across a small range of numbers. The scale and offset can be related to the measured value as follows:

$$\text{Measured Value} = \text{Offset} + \text{Stored\_Value} * \text{Scale\_Factor}$$

For instance, the coldest and hottest surface temperatures on Earth are on the order of 185 K (–88°C) and 331 K (58°C), respectively. The difference between the two extremes is 146 K, so only 146 numbers are needed if you are not using any decimal precision. For example, if I observe a surface temperature of 300 K, I would store this in a two-byte unsigned integer if I do not rescale the data. I can offset our data by 185 K, which is the lowest realistic temperature. Then, I can store this measurement as 115 K, which fits in the one-byte integer ranges. If later I want to read this value, I would then add the 185 K offset back to the value. While reading and writing the data is more complex, the file is now 50% smaller.

I may want further precision, but I can also account for this. For example, at one-digit decimal place (e.g., 300.1 K), the number can be scale multiplied times 10 (which is called the scale factor) and still saved as an integer (Table 3.2). However, our stored value is now 1,515, so this data would be stored as a two-byte integer, which can contain unsigned values up to 65,535. This time, if I am reading the data from the file, value will need to be divided by 10. Again, this conversion saves two bytes of memory over storing 300.15 K as a floating-point value, which is four bytes.

Table 3.2 Examples of How Data Can Be Rescaled to Fit in Integer Ranges

Measured Value	Offset	Scale Factor	Stored Value	Numeric Type
300 K	–	–	300	2-byte integer
300 K	185 K	1.0	115	1-byte integer
300.1 K	185 K	10.0	1,151	2-byte integer
300.15 K	185 K	100.0	11,515	2-byte integer
300.15 K	–	–	300.15 K	4-byte float

Note: Storing data in integers can provide significant storage savings over real values.

If a dataset does rescale the data to improve data storage (and many do not), linear scaling is the most common. Conveniently, some Python packages can automatically detect and apply scale and offset factors when importing certain data types, such as netCDF and HDF formats, which we will discuss in Section 3.2.1.

3.1.2. Arrays

Nearly all satellite data are in three dimensions (3D), which are a value and two corresponding spatial coordinates. The value can, for example, be a physical observation, a numeric code that tells us something about the data, or some ancillary characteristics. The spatial coordinate can be an *x*, *y* or a latitude, longitude pair. Some datasets may also have a vertical coordinate or a time element, further increasing the dimensionality of the data. As an example, I will discuss some different ways you can **structure** (or, how one organizes and manages) three-dimensional data: latitude, longitude, and surface albedo (which shows the reflectivity of the Earth to solar radiation). A simple way to structure the data is in a **list**, an ordered sequence of values. So, I could store the following values:

```
Longitude = [42.7, 42.6, 42.5]
Latitude = [17.5, 17.6, 17.7]
Albedo = [0.30, 0.35, 0.32]
```

Where longitude ranges from –180° to 180° (East is positive), latitude from –90° to 90° (North is positive), and surface albedo from 0 to 1 (which shows low to high sunlight reflection, respectively).

Instead of three separate values, I could organize the data into a **table**, which stores data in rows and columns:

Longitude	Latitude	Albedo
42.7	17.5	0.30
42.6	17.6	0.35
42.5	17.7	0.32

This is helpful because it is human readable, and in a programming sense, I can access the data across the rows rather than work with three different lists. If you are sorting your data, it may also be easier to keep track of the items order in the table form. You may have worked with data in **comma separated variables** (.csv), Google sheets, or Excel format (.xls, .xlsx), which will resemble this structure.

The above two methods are useful if you have a few points, or if you want to plot point observations. However, even the finest satellite data represents an exact point, its value covers an area (or a volume, if there is a vertical component). The total area that the satellite "sees" is known as the satellite **footprint** or **field of view** (FOV).

Rather than use a list, it might then be often easier to store the latitude, longitude values as a multidimensional array (or matrix). For example, values of longitude, latitude, and albedo could be stored in three 2D arrays:

Longitude

42.70	42.68	42.66
42.69	42.68	42.66
42.69	42.68	42.66

Latitude

17.46	17.46	17.46
17.48	17.48	17.48
17.49	17.49	17.49

Albedo

0.30	0.32	0.30
0.35	0.30	0.31
0.34	0.33	0.32

Within Python, this organization is called a **meshgrid**, which stores the spatial and value coordinated in a 2D rectangular grid. Coordinates stored in a meshgrid are either **regularly spaced** or **irregularly spaced data** (Figure 3.2). If every coordinate inside the meshgrid has a consistent distance between their neighboring coordinate, then they are regularly spaced. On the other hand, the irregularly spaced data will have varying distance between the x and y coordinates, or both. In general, if grid coordinates are regularly spaced, the coordinates may be stored in datasets as lists (e.g., GPM IMERG L3). If the data are irregularly spaced, the data will likely be stored as a 2D meshgrid (e.g., in both cases, the data are very commonly stored in a 2D grid).

In Figure 3.3, I show a plot of a single **granule** that contains several **swaths** of surface albedo from 25 June 2018 at 11:21 UTC. Swaths are what a satellite can see

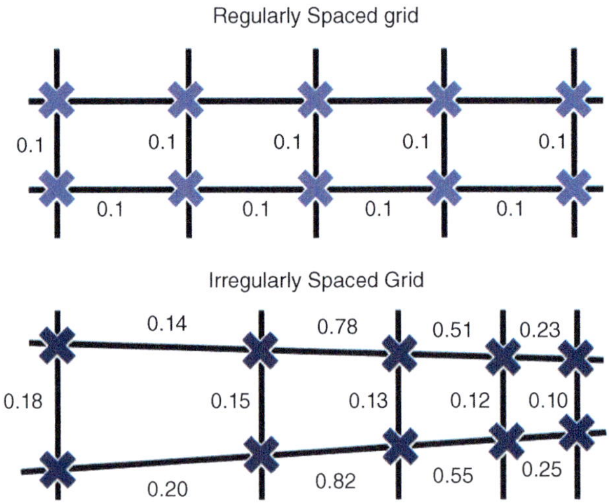

Figure 3.2 Spacing and distance between (*x*, *y*) points for an example regularly and irregularly spaced rectangular grid.

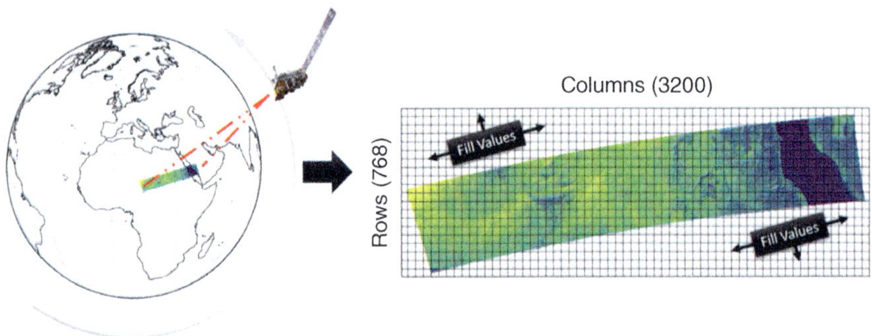

Figure 3.3 An illustration of how a granule of satellite data taken from a polar-orbiting satellite would be stored in a meshgrid. Regions outside of where the satellite can see or that are not stored in the file are indicated using fill values. These values are excluded from analysis.

in longitude and latitude during a scan of the Earth (also called a scanline). A granule is a collection of swaths, usually taken over a time period or for a fixed number of scanlines. Data are often stored in chunks because global scans of the Earth are voluminous. Additionally, the smaller data files improve the **latency** or timeliness of the data because we do not have to wait for the full global image to be assembled.

On the right of Figure 3.3, I illustrate what happens when the data are flattened and projected into a 2D array. Due to the curvature of the Earth and the satellite viewing angle, the coordinates are spatially irregular. At the scan edges, the footprints are larger than they are when the satellite is directly overhead. So, when the data are stored in a rectangular grid, there will be places there is no data. These empty values are called **missing, gaps,** or **fill values**. They will contain a special number that is not used elsewhere in the data, which for continuous data, common values are **NaN** (not a number), –9999.9, –999.9, or –999. These numbers are not particularly common in nature and thus won't usually be confused with meaningful observations, 0 is usually not used as a fill number because it becomes difficult to distinguish low values of observed data with values that are outside of the satellite scan. Data can also contain integers, for example **data flags**, which are integer codes that categorize the data, which may indicate the data quality, or identify if the scene is over land or ocean. For integer data, some common fill values may include 0, 128, –127, 255, and 32,766. These numbers are endpoint values for the numeric data types shown in Table 3.1.

3.2. Data Formats

Data must be combined to provide a complete description of an observed value. Most satellite datasets have a primary **empirical** variable of interest (e.g., precipitation). Then, to display and understand these values, we need **abstract** supporting data such as latitude, longitude, and data quality flags. **Metadata,** such as fill values, helps us understand information about the empirical variable, such where the satellite is taking observations in a granule. In a more general sense, scientific data can be broadly classified into three forms:

- **Empirical data** are observed measurements, including air temperature, precipitation, lightning flash area, or any other measured (or estimated) physical or countable quantity.
- **Abstract data** describe how the empirical data are indexed or stored. For example, latitude, longitude, time, and height are abstract data that describe empirical data like air temperature. Other, nonphysical coordinates are also classified as abstract.
- **Metadata** describes information, formatting, and constraints of empirical data. Metadata includes *attributes* of the data, such as the valid number ranges and if measurements are missing. Metadata can apply to the entire dataset by describing the agency source, the sensor it came from, the algorithm version that created it, and the conventions it uses for storage.

Without well-documented supporting data, even highly accurate empirical data are not useful. As mentioned in the introduction, analysis-ready data must be well-understood and thus would include all the above to ensure data are self-explanatory, efficiently stored, and easily readable. In the following sections, I will describe several data storage methods that are found in the Earth sciences, their strengths and weaknesses, and suitability for analysis and sharing.

3.2.1. Binary

In our daily lives, we use the base-10 numerical. Our counting numbers are 0–9, and then digits increase in length every tenth number. Alternatively, computer data are natively stored in base-2 or **binary**, which takes take on two discrete states, 1 or 0 and can be thought of as "on" or "off." Each state is called a bit. Each additional bit doubles the previous amount of information: one bit allows you to count 0 or 1 (two possible integers), two bits allowed you to count to 3 (four integers), three bits allows you to count to 7 (8 possible integers), and so on (Table 3.3). Note that 8 bits = 1 byte.

Binary data are aggregated into a dataset by structuring bits into sequential arrays. Reading a single-dimension sequence of numbers is relatively simple. For instance, if you know the data type, you can calculate the dimensions of the data by dividing the byte size by the file size. For example, if you have a 320-bit file and you know it contains 32-bit floats, then there it is an array with length 10. However, if you have a multidimensional array, the read order of the programming language will matter. If the 32-but file in the previous example is a 5 × 2 array, then it's not obvious if you are reading the row or the column. Row-major order (used by Python's NumPy, C/C++, and most object-oriented languages) and column-major order (Fortran and many procedural programming languages) are methods for storing multidimensional arrays in memory (Figure 3.4). By convention, most languages that are zero indexed tend to be row major, while languages where the index starts with 1 are column major. If you use the wrong index, the program will read the data in the incorrect order.

The default byte read order (the **endianness**) varies from one operating system or software tool to another; little endian data are read from right to left (most significant data byte last) and big endian is read from left to right (most significant data byte first). You can often tell if you are reading data using the wrong system if you have unexpectedly large or small values.

Table 3.3 Comparing Increasing Numbers in Base-10 to Base-2 (Binary)

Base-10	Binary
0	0000
1	0001
2	0010
3	0011
4	0100
5	0101
6	0110
7	0111
8	1000

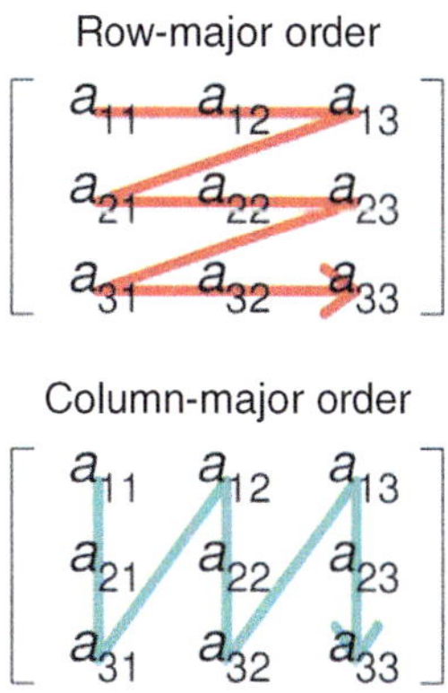

Figure 3.4 Read order for row and column major.

In summary, binary files can be efficiently organized to compactly store empirical data and even abstract data, such as geolocation information. Metadata can also in theory be stored in binary files. However, there is no standard organization or formatting in binary data, so binary data cannot be read unless it includes written documentation describing the internal structure.

3.2.2. Text

Most of us have probably already seen or used text data. Some advantages of text-based data are that they are easily readable, so you can visually check if the imported data matches the input file formatting. However, text data also has different encoding methods and organization. American Standard Code for Information Interchange (ASCII) was released in 1963 and is one of the earliest text encodings for computers. When viewed, text files look like alphanumeric characters, but under the hood ASCII character encoding consists of 128 characters which are represented by 7-bit integers.

Originally, ASCII could only display Latin characters and limited punctuation. ASCII was followed by the Unicode Transformation Format – 8-bit (UTF-8) which produces 1,112,064 valid code points in Unicode using one to four 8-bit bytes. Since 2009, UTF-8 has become the primary encoding of the World Wide Web, with over 60% of webpages encoded in UTF-8 in 2012. Under the hood, UTCF-8 is backward compatible with ASCII data; the first 127 characters in Unicode map to the integers as their ASCII predecessor.

While all languages can be displayed in UTF-8, UTF-16, and UTF-32 are becoming more frequently used for languages with more complex alphabets, such as Chinese and Arabic. Many scientists will usually not notice these distinctions and may use the terms interchangeably. However, knowing the distinction is important, as using the wrong format can introduce unwanted character artifacts that can cause code to erroneously read files.

Of plain text data formats, comma-separated values (.csv files) and tab-separated values (.tsv files) are highly useful for storing tabular data. Below is an example of a csv file:

```
STATION_NUM, STATION_NAME, LATITUDE, LONGITUDE,
ELEVATION USW00014768,ROCHESTER, NY, 43.1167, -77.6767,
164.3
USC00309049,WEBSTER, NY, 43.2419, -77.3882, 83.8
```

An example of a tsv dataset is the following METeorological Aerodrome Reports (METAR), which is a text-based format used by pilots for aviation planning.

```
KBWI 291654Z VRB04KT 10SM FEW045 27/13 A2999 RMK AO2
SLP155 T02720133
```

Note that just because text data are "human readable," does not mean it is "human understandable" without additional explanation. However, the above code is very "machine readable" because it follows consistent formatting and uses a common set of codes to describe the current weather conditions. For instance, the above METAR can be translated from "machine readable" into a "human-readable" description using a text-parsing algorithm written in Python:

```
Location: BWI Airport
Date: August 29, 2019
Time: 16:54Z
Winds: variable direction winds at 4 knots
Visibility: 10 or more sm
Clouds: few clouds at 4500 feet
AGL Ceiling: at least 12,000 feet
AGL Pressure (altimeter): 29.99 inches Hg
Sea level pressure: 1015.5 mb
Temperature: 27.20°C
Dewpoint: 13.3°C
```

It is somewhat arguable that METARs in the first form are a "user-friendly aviation weather text product" as described on the official aviation weather webpage. But to make it human-understandable, the above code can be easily imported and parsed by Python. In this way, a user can write a function to convert the data to make it understandable to a layperson.

Like binary data, text files can store empirical and abstract data, and some simple metadata. However, there are some disadvantages of character code. For one, the data require much more storage space than binary data formats, which do not require as many bytes to store the data. Second, formatting csv and tsv files needs to be carefully organized to be machine readable: column headers may not match the number of columns below, the separating or delimiting character may be inconsistent, or there may be missing values. Third, sometimes multiple variables are stacked into the same file, rather than organized by rows, and a special reader must be designed to import the data.

In the next section, I will discuss self-describing datasets, which in more recent years have become the primary method of storing large satellite datasets. Self-describing datasets combine the compactness of binary data with the human-readable descriptions of text files so that users can understand the structure, metadata, and formatting of the stored empirical and abstract data.

3.2.3. Self-Describing Data Formats

Self-describing formats are the main vector of communicating satellite data. In terms of their design, they are a collection of binary data organized in a standardized way. Because multiple datasets are easily stored in a single file, you can open the file and learn which empirical variables are available, the time of the overpass, data quality information, and the geolocation information if the data are gridded. With these descriptors, the data are ready to use off the shelf. For instance, if you want to plot aerosol optical depth (AOD), you will be able to see exactly where in the file the data are located as well as some other useful information like the units of the data. If you are comfortable navigating directories on your laptop, you will find the layout of the binary files similar.

Self-describing formats are powerful, but I admit that they can be incredibly challenging to learn initially. However, the reward is that your knowledge is transferrable to many satellite datasets, not to just to one. The increased use of standardized self-describing formats has helped Earth science dramatically where consistent formats have made it easier to disseminate data through centralized data portals. In contrast to the Earth sciences, the space sciences community continues to struggle with a myriad of formats and access points and where previously each product developer would use any format or structure of their choosing – a hurdle for early career scientists.

What are the advantages of self-describing binary data over plain text ASCII data that make this additional complexity worthwhile? The reason is primarily economical – both in a monetary sense and a computational one. On the NASA Earthdata portal alone, there are over 850 datasets that extend back to the beginning of the satellite era. As discussed at the beginning of this chapter, early records were stored on hard drives, which over time degraded in warehouses (James et al., 2018). Compact formats enable agencies to retain longer records. Binary-based formats take up less space on a storage disk than text data. Each byte of binary data can hold 256 possible values whereas **unextended text** (text that is exclusively machine readable) can only store 128 and human readable text is even less efficient. Computationally, it is faster to read in binary-based datasets than text, which needs to be parsed before being stored into a computer's memory. Because the files are more compact, binary formats are commonly used to store large, long-term satellite data.

The most common self-describing format is the Hierarchical Data Format (HDF5). HDF files are often used outside of the Earth sciences, as they are useful for storing other big datasets. The next common format is Network common data format (netCDF4), which is derived from HDF. Standards for netCDF are hosted by the Unidata program at the University Corporation for Atmospheric Research (UCAR). As a result, these files are very often found in the Earth sciences. NOAA remote sensing datasets are almost exclusively netCDF, and more recently, NASA produced datasets have been more frequently stored in netCDF4. Please note that older versions of the datasets (HDF4, netCDF3) do exist but are not necessarily backward compatible with the Python techniques presented in this book.

Before getting started with any data-based coding project, it is a good practice to inspect the data you are about to work with. Self-describing datasets are a type of structured binary format. While the **variable** data itself is binary, it is organized into **groups**. I recommend that after opening the dataset, you examine which specific elements within it are of interest to you. Additionally, self-describing datasets can have both local and global **attributes**, such as a text description of the variable, the valid range of values, or the number that is used to indicate missing or fill, values (Figure 3.5). In self-describing formats you do have to know where in the large array your variable of interest is stored. You can extract it by using the variable name, which points to the address inside of the file. Because the data can be complex, it is worthwhile to inspect the contents of a new dataset in advance. You can utilize free data viewers to inspect the dataset contents, such as **Panoply** (NASA 2021). Python or command line tools also allow you to examine what is inside the files. Planning your coding project with knowledge of the data will allow you to work efficiently and overall will save you time.

The variables can be one-, two-, three-dimensional variables, or more. HDF files may organize the variables into **groups** and **subgroups**. While variables can be organized into groups in netCDF files, Unidata compliance standards do not recommended grouping. Most often, data can be separated into other variables. If the data have the same grid and there is a temporal element, it is more efficient to increase the dimensionality of the variables and stack the matrices in time.

3.2.4. Table-Driven Formats

Binary data that take on **table-driven code form** require external tables to decode the data type. Thus, they are not self-describing. These files follow a methodology of encoding binary data and not a distinct file type. Binary Universal Form for the Representation of meteorological data (BUFR) and GRIdded

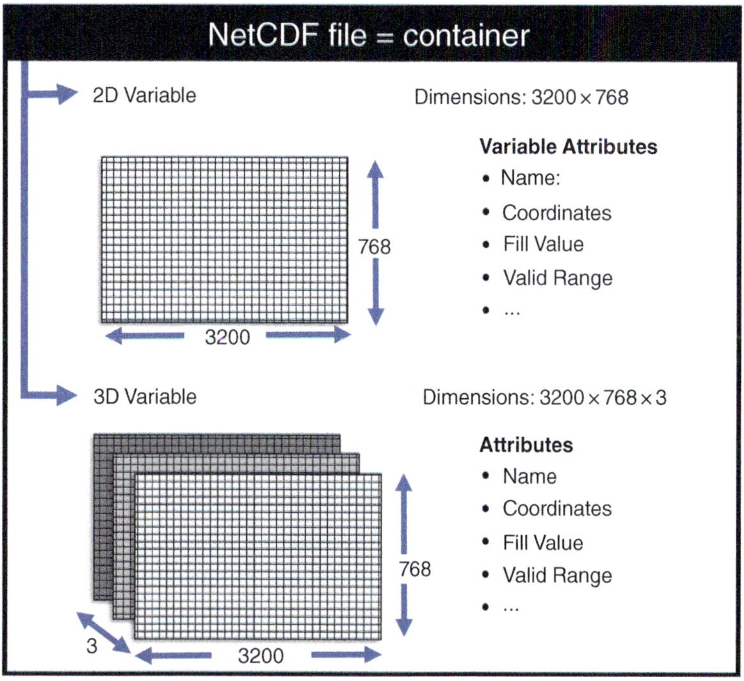

Figure 3.5 Example of how netCDF data are organized. Each variable has metadata that stores units, fill values, and a more detailed description of the contents.

Binary (GRIB) are two common table-driven formats in Earth Sciences, but they are specific to certain subject areas. I will not cover these formats in significant detail in this text but will mention them here for your awareness.

- **BUFR.** BUFR was developed by the World Meteorological Organization (WMO). Many assimilated datasets are stored in this format. BUFR files are organized into 16-bit chunks called messages, with the first bits representing the numeric data type and format codes followed by the bit-stream of the data. The software that parses these files uses external tables to look up the meaning of the numeric codes, with the WMO tables as the field standard. BUFR Table A is known as the Data Category. If the Data Category bit stream contains the number 21, the Python parser can use the tables to see that this code corresponds to "Radiances (satellite measured)." An advantage of BUFR is that the descriptions of the data are harmonized and has superior compression to text-based formats. When BUFR files follow standards, they are more easily decoded than plain binary files. However, if the stored data does not conform to the codes in the WMO tables, then the data must be distributed with a decoding table (Caron, 2011).

- **GRIB2.** American NWS models (e.g., GFS, NAM, and HRRR) and the European (e.g., ECMWF) models are stored in GRIB2. While they share the same format, there are some differences in how each organization stores its data. Like BUFR, GRIB2 are stored as binary variables with a header describing the data stored followed by the variable values. GRIB2 replaced the legacy GRIB1 format as of 2001 and has a different encoding structure (WMO, 2001).

Models merit discussion in this text because researchers often compare models and satellite data. For example, models are useful for validating satellite datasets or supplementing observations with variables that are not easily retrieved (e.g., wind speed).

At the time this text was published, many Python readers had been developed and tested with ECMWF because historically, most Python developers have been in Europe. For instance, some of the GRIB2 decoders have problems parsing the American datasets because the American models have multiple pressure dimensions (depending on the variable) while the European models have one. Still, there are ways the data can be inspected by using the **pygrib** and **cfgrib** packages, which were described later in Section 2.4.

3.2.5. geoTIFF

GeoTIFF is essentially a geolocated image file. Like other image files, the data are organized in regularly spaced grids (also known as raster data). In addition, GeoTIFF contains metadata that describes the location, coordinate system, image projection, and data values. An advantage of this data format is that satellite imagery is stored as an image, which can be easily previewed using image software. However, like text data, the data are not compact in terms of storage. While gaining popularity in several fields, geoTIFF is most used for GIS applications.

3.3. Data Usage

There is a lot of technical jargon associated with satellite data products. However, having a working knowledge of the terminology will enable you to understand how to appropriately choose what data to use. Below, I discuss how several data producers (e.g., NASA, NOAA, ESA) define the processing levels, the level of data maturity, and quality control. The timeliness of the data may also be of importance, so we describe what is meant by data latency. Finally, the algorithms used to calibrate and retrieve data change over time, so it is important to be aware of what version you are using for your research.

3.3.1. Processing Levels

Data originating from NASA, NOAA, and ESA are often assigned a processing level, numbered from 0 to 4 to help differentiate how processed the data are (Earthdata, 2019; ESA, 2007). Specific definitions can vary between agencies, but in general, Level 0 is raw, unprocessed satellite instrument data that is typically encoded in binary at full resolution and is not often used for research or visualization. Instead, Level 0 data are used inside data production environments, which process the raw data from a binary format to understood units such as radiance, bending angles, and phase delays (Level 1 data). Level 1 data are still minimally processed (e.g., converted to an HDF or BUFR) and is further divided into three categories (A, B, or C) with some subtle distinctions. **Level 1A** data may contain the correction and calibration information in the dataset, but it is not applied to data. Calibration ensures that the raw measurements are consistent and meet specific requirements for scientific research and numerical weather prediction. **Level 1B** has calibration applied but is not quality controlled and is not spatially or tem porally resamples. **Level 1C** can be resampled, calibrated, or corrected, and may also include quality control (Section 3.2.3). **Level 2** data are geophysical variables, such as precipitation rate, AOD, surface type, or insolation. Conversion from Level 1 to Level 2 is not straightforward, in part because raw signals such as electromagnetic radiation are often only indirectly linked to the desired variable. Furthermore, there are not always enough channels to obtain unique solutions without further assumptions or constrains. **Retrieval algorithms** are developed to process Level 1 data along with ancillary datasets to make a reasonable estimation of geophysical variables such as temperature, atmospheric composition, or surface type. Levels 1 and 2 data are often stored in a granule, which is a combination of all observation within a given time period (e.g., 1 minute of data). In terms of data management, there will be many relatively small files. **Level 3** data are spatially and temporally resampled geophysical variables from Level 2 data. For instance, Level 3 data may combine Level 2 data from multiple sensors to create a global snapshot or aggregate instantaneous measurements to produce monthly mean values. **Level 4** data synthesizes Level 1–3 satellite observations with models to create a consistent and often global, long-term reanalysis of the geophysical state. Depending on aggregation and number of available variables, Level 3 and 4 files may be significantly larger than a single Level 2 file. Table 3.4 shows a short summary with a few sample products that fall into the categories.

Table 3.4 Levels and Examples of Transformations Performed on the Data

Level	Processing	Sample Dataset(s)
0	Raw data	N/A
1	Calibrated or uncalibrated radiances or brightness temperatures	VIIRS or ABI radiance data
2	Products, converted to estimate a physical quantity from Level 1 data	Dark Target AOD Product
3	Combination of multiple Level 2 data, daily or monthly average	Hourly CMORPH-2, Monthly NDVI
4	Combination of remote sensing, in-situ, or model data	Merra-2 Global Reanalysis

Generally, Level 3 and Level 4 data may be more appropriate to study long-term trends or teleconnections. For real-time environmental monitoring or region-specific analysis, Level 2 datasets may be more useful. Modelers may be interested in assimilating radiances from Level 1 data, and on occasion Level 2 data, such as for trace gas assimilation.

3.3.2. Product Maturity

When new NASA or NOAA satellites are launched, there is a test period where the products are evaluated, which is sometimes referred to as the "alpha" or "post-launch" phase. After this test period, so long as the data meet minimum standards, the data are in the beta stage. **Beta** products can sometimes restricted to the public, although access may be requested by users from the agencies under specific circumstances. The next two stages are **provisional** and **validated**. At these maturity stages, you can utilize these datasets with more confidence. Both provisional and validated datasets are appropriate for scientific research.

In terms of appropriate use, data that have beta maturity or below really are not scientifically rigorous. This data may still contain significant errors in the values and even geolocation. There might be certain applications where using beta stage data are appropriate, such as if you want to do a feasibility study, want to simply become familiar with the data, or if you are evaluating the data accuracy and precision. If these data are used in published scientific research, you should clearly state the maturity level in your manuscript or presentation.

In terms of time scales, advancement through maturity stages is slower for brand-new sensor, anywhere about one to two years from launch. Level 1 products are always the first to become validated – they have fewer moving parts and produce the simplest of measurements. Level 2 products can at times rely on other Level 2 products as inputs. For instance, many of the GOES-16 products rely on the Level 2 Clear Sky Mask to determine if a scene is clear or cloudy. So, the Clear Sky Mask had to reach provisional before other products could. Logistically, this can complicate the review process for downstream products. For a series of satellites, the review process is expedited. For instance, GOES-17 was a twin satellite to GOES-16, so the review process for GOES-17 products was faster than for GOES-16 products.

3.3.3. Quality Control

A sensor's ability to retrieve an atmospheric variable is strongly influenced by underlying atmospheric and surface conditions. For example, instruments that measure visible or IR cannot penetrate cloud tops, so the "best" surface observations occur over cloud-free scenes. Bright background surfaces like snow, desert sand, or sun glint on the ocean can also impact the accuracy of the retrieval. Satellite datasets often include quality control flags to inform users if and where the dataset contains degraded or unphysical observations.

There is no standard for communicating quality control across different retrievals. This is because retrieval algorithms are complex and the needs of the user community are diverse. Some datasets will use a binary flag ranked as good (0) or bad (1), while other datasets reverse the numbers. Others will use categorical labels such as high, medium, or low quality or a continuous scale, 0–100%. Some do not use a single measure but rather combine many bit flags, which can indicate data quality according to specific criteria.

For robust scientific research, you will almost always want to use the best-quality data. However, this, in turn, reduces the number of available observations for your study. In some cases, it might be reasonable to keep the top-two quality flags (high and medium) and only discard the lowest-quality data. There are situations where examining all available data are valuable, such as in data sparse regions where no other measures are available.

3.3.4. Data Latency

For many scientists, it is important to know when to access data as much as where and how to access the data. Most research scientists are content with accessing data days, months, or years after the observations are made. Often the resulting work is built on long-term means or case studies of a past event. However, those involved with hazard management, such as flood or severe weather forecasting, will need to access the data sooner. Data that are provided immediately after downlinking from the satellite is known as **near real-time data**.

As described in Section 1.3, Level 0 data are downloaded from the satellite to ground downlink stations. From there, antennas distribute the observations to data processing centers around the world. Level 1 data are often released first, since they have the least amount of processing, and often no dependency on other products, and are prioritized for assimilation into forecasting and climate models. Level 2 and Level 3 datasets will take longer to be released; they require more products as inputs and may call computationally expensive routines such as radiative transfer models. Recall from Section 1.3, data latency is the time difference between when a satellite makes an observation from when the data are made available to the end user. The latency at which a user receives the data is determined by how often a satellite is downlinked to the Earth, the speed of the retrieval algorithm, bandwidth, and the needs of data users. To decrease latency over the United States, observations from some polar orbiting satellites are downloaded by a network direct broadcast receivers. Retrievals are then processed locally using the free, open-source Community Satellite Processing Package (CSPP) developed by the University of Wisconsin. Each individual site can then internally use (or share via FTP) the data at a faster pace. Figure 3.6 shows the antenna sites across the United States and Pacific.

A disadvantage of direct broadcast is that the receiver can only download the local data from polar orbiting satellites during the short time that they are in range.

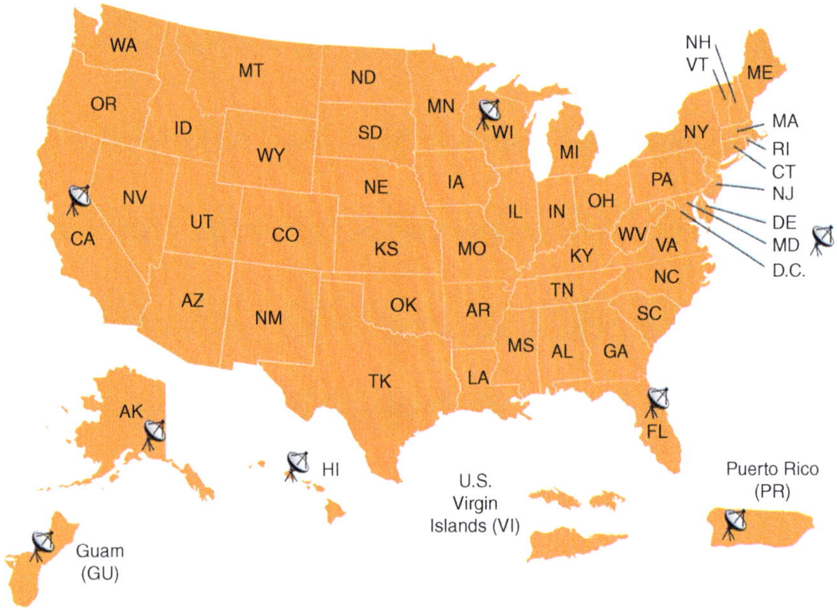

Figure 3.6 Direct broadcast (DB) antenna sites, which can provide data in real-time. Source: Joint Polar Satellite System/Public Domain.

Data from multiple receivers can be combined into a larger, regional image. However, global view of the data is only possible at two satellite ground stations located in Svalbard, Norway, and Troll, Antarctica.

IMERG, a global Level 3 precipitation product that combines many passive microwave sensors from various platforms, is critical for flood monitoring in remote regions without access to radar. However, the logistics of combining observations and running the retrieval make distribution challenging. The solution that was adopted was to produce three versions of the dataset:

- Early run: Produced with a latency of 4 hours from the time all data are acquired.
- Late run: Produced with a latency of 12 hours.
- Final run: Research grade product that is calibrated with rain gauge data to improve the estimate. The latency is 3.5 months.

The required timeliness of datasets varies between different agencies, missions, and priorities. NASA datasets tend to focus on long-term availability and consistency of the data to create climate records. In contrast, NOAA datasets tend to focus on capturing the state of the atmosphere and oceans in the present to serve operational forecasting. So, you will likely find more real-time datasets generated by forecast agencies such as NOAA.

3.3.5. Reprocessing

Over time, research leads to improvements in the quality of retrieval algorithms. For the data distributors, they may choose to simply make the switch and not change prior data, or they may choose to reprocess what is often years of data to make the entire record consistent. NOAA data are seldom reprocessed because the organization prioritizes forecasts and warnings. NASA records tend to be available for longer periods of time and are generated using a common retrieval algorithm. To ensure that you are working with the same version of the algorithm, this information is typically stored in the global metadata of the HDF or netCDF files.

3.4. Summary

In this chapter, you learned some of the main scientific data types and common terminology. In the past, satellite data was distributed as text or binary datasets, but now netCDF and HDF files are more common. These self-describing data are advantageous because they contain descriptive metadata and important ancillary information such as georeferencing or multiple variables on the same grid. There is a significant amount of remote sensing data, so it can be overwhelming for new users to know which data are appropriate to use for research or environmental monitoring. Scientific data from major international agencies often classify them according to processing level to convey information on the degree of calibration, aggregation, or combination. Furthermore, quality control flags within the datasets can help discriminate which measurements should and should not be used.

Because no two satellite datasets are identical, universally understood formats such as netCDF or HDF make it easier for scientists to compare, combine, and analyze different datasets. Since there is ongoing research to improve retrieval algorithms, self-describing formats can also include production information on the version number and maturity level of the data from new satellite missions. Overall, unification of datasets, language, and production promotes analysis-ready data (ARD) so that scientists can more easily use tools like Python for research and monitoring.

References

Balaraman, Goutham. (2017, March 14). Numpy vs Pandas performance comparison. Retrieved April 5, 2020, from http://gouthamanbalaraman.com/blog/numpy-vs-pandas-comparison.html

Caron, J. (2011). On the suitability of BUFR and GRIB for archiving data. Unidata/UCAR. Retrieved from https://doi.org/10.5065/vkan-dp10.

Dwyer, J. L., Roy, D. P., Sauer, B., Jenkerson, C. B., Zhang, H. K., & Lymburner, L. (2018). Analysis ready data: Enabling analysis of the Landsat archive. *Remote Sensing, 10*(9), 1363. https://doi.org/10.3390/rs10091363

Earthdata (2019). Data processing levels. Retrieved April 18, 2020, from https://earthdata.nasa.gov/collaborate/open-data-services-and-software/data-information-policy/data-levels/

ESA (2007, January 30). GMES Sentinel-2 mission requirements. European Space Agency. Retrieved from https://www.eumetsat.int/website/home/Data/TechnicalDocuments/index.html

James, N., Behnke, J., Johnson, J. E., Zamkoff, E. B., Esfandiari, A. E., Al-Jazrawi, A. F., et al. (2018). NASA Earth Science Data Rescue Efforts. Presented at the AGU Fall Meeting 2018, AGU. Retrieved from https://agu.confex.com/agu/fm18/meetingapp.cgi/Paper/381031

Meier, W. N., Gallaher, D., and G. C. Campbell (2013). New estimates of Arctic and Antarctic sea ice extent during September 1964 from recovered Nimbus I satellite imagery. The Cryosphere, 7, 699–705, doi:10.5194/tc-7-699-2013.

Murphy, K. (2020). NASA Earth Science Data: Yours to use, fully and without restrictions | Earthdata. Retrieved April 17, 2020, from https://earthdata.nasa.gov/learn/articles/tools-and-technology-articles/nasa-data-policy/

NASA (2021). Panoply, netCDF, HDF, and GRIB data viewer. https://www.giss.nasa.gov/tools/panoply/

Precipitation Processing System (PPS) At NASA GSFC. (2019). GPM IMERG early precipitation L3 half hourly 0.1 degree x 0.1 degree V06 [Data set]. NASA Goddard Earth Sciences Data and Information Services Center. https://doi.org/10.5067/GPM/IMERG/3B-HH-E/06

WMO (2003, June). Introduction to GRIB Edition1 and GRIB Edition 2. Retrieved from https://www.wmo.int/pages/prog/www/WMOCodes/Guides/GRIB/Introduction_-GRIB1-GRIB2.pdf

Special thanks to Christopher Barnet for his description of historic plotting routines.

Part II

Practical Python Tutorials for Remote Sensing

4

PRACTICAL PYTHON SYNTAX

Python is an excellent language for both new and novice programmers. This chapter illustrates basic Python skills through a series of short tutorials, such as using Jupyter Notebooks, assigning variables, creating lists and arrays, importing packages, working with masked arrays, and creating dictionaries. This chapter also covers loops, logic, and functions. These skills are highlighted and will be used again elsewhere in this text.

In this chapter, you will take your first steps in Python. If you are brand new to programming, this will not likely be enough to get fully up to speed. You can start by working through the exercises but if I move too quickly, you may have to supplement your learning with another resource (I have some recommendations in Appendix C). If you have experience writing another programming language, this chapter will provide a basic overview that will be helpful in later chapters.

Python code can be written in either object-oriented (code is organized to store data and functions as objects) or procedural-oriented paradigms (code is organized as a series of steps and function calls). You do not need to be familiar with these terms to use this book. Some of the methods and operations I use in this book follow the object-oriented paradigm, and some of the code may follow a procedural paradigm. Since my goal is to help you perform common tasks using Earth observations, I am designing the example code to help you understand how to perform some basic methods, applications, and visualizations. To keep the discussion focused, I will not describe computer science philosophy or software engineering techniques. As you progress on your scientific programming journey, you will naturally want to learn more about these topics or explore Python's many capabilities, which I will briefly cover in Chapter 11 and 12.

Earth Observation Using Python: A Practical Programming Guide, Special Publications 75, First Edition. Rebekah B. Esmaili.

DOI: 10.1002/9781119606925.ch4

But for now, please feel comfortable to learn and explore. All code examples are downloadable and can be run interactively online: https://resmaili.github.io/Earth-Obs-Py. This code is yours – make any changes you wish and let your curiosity drive your learning.

If you have not yet installed Python and **Jupyter Notebook**, visit Appendix A and B to get setup. Notebooks are a valuable tool for learning because they allow you to learn in an interactive mode, which allows you to write code blocks, execute them, and immediately print the results. Also note that I will use **Python 3** for all examples in this book. I strongly advise not writing any new code in Python 2.7, as it is no longer being maintained by the community.

4.1. "Hello Earth" in Python

The most basic line command is to write words to the screen, so below I will print "Hello Earth" to our screen. To do this, I must give Python a **command**, which is an instruction that Python interprets and performs a task. In the example below and throughout the text, the code is on top while the output is italicized and the last line of text in each code block.

```
print("Hello Earth")
Hello Earth
```

To run the above command in Jupyter notebook, highlight the cell and either click the run button (▶) or press the `Shift` and `Enter` keys. If successful, the words within the `print` statement will appear below the line of text. If this was your first-time running Python, congratulations! This may seem like a small start, but in printing the above statement, your installation and environment is set up and you have even learned to run code inline.

Try altering the above statement, removing a quote or parenthesis, to examine what an error message looks like. In Jupyter Notebook, when you do not see an error message inline, it means the code ran successfully. Python error messages can be verbose; it is often useful to begin by examining the last thing printed and working upward.

4.2. Variable Assignment and Arithmetic

Variables are used by programs to store information such as numbers or character strings of text. In some older programming languages, the user had to explicitly tell the program what the variable type and size was, such as four-byte float, two-byte integer (both are gray in code blocks) or 12-character string (red text in code blocks). In Python, variables are dynamically allocated, which means that you do not need to declare the type or size prior to storing data in them. Instead, Python will automatically guess the variable type based on the content of what you are assigning. Note that since we are only assigning values to variables, no output is printed:

```
var_int = 8
var_float = 15.0
var_scifloat = 4e8
var_complex = complex(4, 2)
var_greetings = 'Hello Earth'
```

Functions are a type of command that contains groups of code statements that are designed to perform a specific task. To use a function, the syntax is `function_name(input)`. You can write your own functions, but there are many **built-in** Python functions (these are green in the example code blocks in this book). So far, I have used two function: `print` and `complex`. Another useful function is `type`, which will tell us if the variable is an integer, a float, or a string. If you are unfamiliar with data types, refer back to Section 3.2.1.

```
type(var_int),  type(var_float), type(var_scifloat),
type(var_complex), type(var_greetings)

(int, float, float, complex, str)
```

Even without the `print` function, Jupyter Notebook will automatically write the **result** of the last command in the code block, if there is one. So, I typically omit print statements unless it improves readability of the output.

Occasionally, I will intentionally or accidentally perform mixed-mode operations, where I combine integers and floats (strings cannot be combined with either, doing so will return an error). Note that if I do not use print(), Jupyter Notebooks will automatically print the last line of unassigned code:

```
var_float/var_int
1.875
```

Multiplication, addition, and subtraction operators on integers return integers. However, for division operations, Python 3 performs an automatic type conversion from integer to float ("true division"). This behavior is unlike IDL, Fortran, and Python 2.7, which would return 0 as the solution for a mixed mode calculation ("classic division"). However, if you want the result as an integer, then you would need to convert the solution back to an integer:

```
int(var_float/var_int)
1
```

You can also convert between string and numeric:

```
var_string = '815'
int(var_string)
815
```

There are some caveats. Following IEEE 754 standards, the largest number you can use in Python must be less than 1.8×10^{308}. Anything larger is infinity. Also, Python will round anything less than 5.0×10^{-324} to zero. If you are familiar with numerical analysis, you probably realize that Python floats are signed 64-bit floats.

Python has the standard assortment of built-in mathematical operations that can be performed: addition, subtraction, division, multiplication, and exponentiation. Anything built-in in Fortran and C++ is available to you in Python. Mathematical operation symbols are highlighted in pink in the example code blocks.

```
add_sub = 8+15-42
mult_div_exp = (8*15/42)**4
print(add_sub, mult_div_exp)
-19 66.63890045814244
```

The quotient (integer division) and modulo (remainder from the division) are also built into standard Python:

```
div_quo = 15 // 8
div_mod = 15 % 8
print(div_quo, div_mod)
1 7
```

There are also numerous functions in standard Python. For instance, you can round floating-point numbers (`round`) and get the absolute values (`abs`). Table 4.1 shows some useful functions and methods that I will use in this text. The terms *methods* and *functions* both refer to callable subprograms that return values, but methods are often associated with object-oriented programming. The difference may not be entirely transparent. In general, a method will be called using `variable.min()`, for instance, instead of the `min(`*`variable`*`)` function syntax.

```
abs(add_sub), round(mult_div_exp)
(19, 67)
```

Table 4.1 Useful Python Functions and Methods Used in This Text

abs()	all()	any()	bin()
bytearray()	bytes()	callable()	dict()
dir()	divmod()	enumerate()	eval()
filter()	float()	format()	int()
iter()	len()	list()	max()
min()	print()	range()	reversed()
round()	sorted()	str()	sum()
tuple()	type()		

4.3. Lists

Scientists more often work with data, not single values, and **lists** and **arrays** are useful for storing the data. Lists are made using square brackets. They can hold any data type (integers, floats, and strings) and even mixtures of the two:

```
mixed_list = [8, 15, "Oceanic"]
mixed_list
  [8, 15, 'Oceanic']
```

You can access elements of a list using their index. Note that like C/C++ and IDL, Python indices are zero based. So, to access the first element, the index is 0; for the second element, the index is 1, and so on:

```
mixed_list [0], mixed_list [2]
 (8, 'Oceanic')
```

New items can also be appended to the list, which is updated in place. In the example below, the syntax follows `variable.function(arguments)`:

```
mixed_list.append("Flight")
mixed_list
  [8, 15, 'Oceanic', 'Flight']
```

However, you cannot perform mathematical operations on the list. For example, if you try to multiply a list of numbers by the integer 2 using $*$, instead of the amount of each element being doubled, the integers are repeated. This operation doubles the size of the list:

```
numbers_list = [4, 8, 15, 16, 23, 42]
numbers_list*2
 [4, 8, 15, 16, 23, 42, 4, 8, 15, 16, 23, 42]
```

If you instead multiply by the float 2.0, Python will return an error message. Subtracting and division operations are also unsupported operands, and using them returns errors. "Addition" also works, but rather than adding terms, it will append two lists (also called **string concatenation**). For instance:

```
numbers_list + mixed_list
[4, 8, 15, 16, 23, 42, 8, 15, 'Oceanic', 'Flight']
```

I can't easily perform iterative mathematical calculations on lists, but I can perform them on arrays. To do this, I will have to import a package into the notebook.

4.4. Importing Packages

Packages are collections of modules that help simplify common tasks. I went over some important packages in Section 2.3, but so far I am working in base Python and have not linked any of the packages into this particular notebook. Stylistically, this is typically done at the top of the script or notebook, although the libraries can be called at any point in the code. For learning purposes, I am calling the packages as I introduce them, but I recommend you put them at the top of any code you develop independently.

So, let's import a package, **NumPy**, which is useful for mathematical operations and array manipulation. I can run the previous example and get the results I was expecting.

The basic syntax for calling packages is to type `import [package name]`. However, some packages have long names, so I can use `import [package name] as [alias]`, like I show below:

```
import numpy as np
```

If you have this package installed, then running this code block will not return anything to your screen. However, if you do not have this package installed, an error will be returned (See Appendix A, Package Management for an example). So, if nothing happened after you ran the code above, that is good news!

Using the full name (`numpy`) or an alias (which can be thought of as a "nickname") makes it easier to track where the functions you are calling originate. On occasion, you may only need one module, function, or constant from a package. In this case, I lead with the `from` statement:

```
from numpy import pi
print(pi)
3.141592653589793
```

However, I do not recommend the following:

```
from numpy import *
```

If you use `import *`, then it is harder to track which package the functions are being called from. Additionally, packages can have functions of the same name (e.g. `pi` is available in both the math package and NumPy). Python will use the function from the package that was imported last, but you may not be aware of this conflict (Slatkin, 2015). So, in the spirit of transparent and reusable code, it's better to avoid the second command above.

4.5. Array and Matrix Operations

Now that I have imported NumPy, I can use NumPy's array constructor (`np.array`) to convert our list to a NumPy array and perform the matrix multiplication.

```
numbers_array = np.array(numbers_list)
numbers_array*2
array([ 8, 16, 30, 32, 46, 84])
```

I can access elements of the array using indices:

```
numbers_array[0]
4
```

I can rearrange the numbers in sequential order using `sort`:

```
np.sort(numbers_array)
array([ 4,  8, 15, 16, 23, 42])
```

I can also see that the multiplication above did not overwrite the original array. If you wanted to keep the results, the array would have to be assigned to a new variable or overwrite the existing one.

Lists are only one-dimensional. NumPy can be any number of dimensions:

```
numbers_array_2d = numbers_array.reshape(2,3)
numbers_array_2d
array([[ 4,  8, 15],
       [16, 23, 42]])
```

I use indices to access individual array elements. The syntax for a two-dimensional array is [row, column]. There are some caveats: If indices are out of bounds, NumPy will throw an error and stop. Additionally, if you pass NumPy a negative integer, it will "wrap around" and retrieve the last element (if passed –1), second to last element (if passed –2), and so on.

```
numbers_array_2d[1,1]
23
```

Typing arrays can get tiresome, but fortunately, there are some helpful functions that can enable to us to generate them faster. For instance, you can create an empty array with zeroes, NaN values, or a sequence of numbers using the following functions:

```
np.zeros(10), np.full(10, np.nan), np.arange(0, 5, 0.5)
(array([ 0., 0., 0., 0., 0., 0., 0., 0., 0., 0.]),
 array([ nan, nan, nan, nan, nan, nan, nan, nan, nan, nan]),
 array([ 0. , 0.5, 1. , 1.5, 2. , 2.5, 3. , 3.5, 4. , 4.5]))
```

As you can see above, `np.zeros` makes a list of 10 values, all equal to zero. `np.full` prints any value, and in the example I provided, NaN is the fill value. `np.arange(0, 5, 0.5)` will print a list of values from 0 to 5 with an increment by 0.5, following start ≤ x < end. Thus, the end value (5.0) is excluded. To include the end value, you will need to increase the end number by some amount (e.g. 5.1 instead of 5.0). If the step (0.5) was not provided, the default increment is 1.

When working with data, sometimes there are numbers I want to remove. For instance, I may want to work with data below a certain threshold. I can manually remove the data by creating a mask using Boolean operators, as shown in Table 4.2.

Now, I can find which elements of the array that meet some condition, such as

```
masked_nums = (numbers_array > 8)
masked_nums
array([False, False, True, True, True, True], dtype=bool)
```

This returns an array of True and False values if the condition was met. This is also confirmed with the `dtype` attribute at the end of the printed array. I learned that I can subset arrays or access specific values by passing an integer index into the array. I can also pass true and false values to access elements I want (`True`) and remove those I do not (`False`). I can use the `masked_nums` array to subset `numbers_array`:

```
numbers_array[masked_nums]
array([15, 16, 23, 42])
```

Table 4.2 Built-in Comparison, Identity, Logical, and Membership Operators in Python

<table>
<tr><th>Operation</th><th>Meaning</th></tr>
<tr><td><</td><td>less than</td></tr>
<tr><td><=</td><td>less than or equal to</td></tr>
<tr><td>></td><td>greater than</td></tr>
<tr><td>>=</td><td>greater than or equal to</td></tr>
<tr><td><=</td><td>greater than or equal to</td></tr>
<tr><td>==</td><td>equal</td></tr>
<tr><td>!=</td><td>not equal</td></tr>
<tr><td>is</td><td>object identity</td></tr>
<tr><td>is not</td><td>not object identity</td></tr>
<tr><td>&, and</td><td>and</td></tr>
<tr><td>|, or</td><td>or</td></tr>
<tr><td>not</td><td>not</td></tr>
<tr><td>in</td><td>in</td></tr>
<tr><td>not in</td><td>not in</td></tr>
</table>

Notice that the original array had six elements and returned array has four. There are cases where I want to preserve the size and shape of the original array. For these situations, I can utilize the NumPy masked array module. I can convert lists and arrays to this as a masked array. However, I will add a keyword argument `mask=`, which is set to the inverse (~) of the `mask_nums`. True values will become false, and vice versa:

```
numbers_mask_array = np.ma.array(numbers_array,
     mask=~masked_nums)
numbers_mask_array, numbers_array.size
```

```
(masked_array(data = [-- -- 15 16 23 42],
        mask = [ True  True False False False False],
   fill_value = 999999), 6)
```

The returned object contains the source data, an array called the mask, and a fill value – which by default is 999999. You can change this value to whatever you like by using the keyword argument `fill_value` when you define the array.

Also notice that unlike the example using regular arrays, the masked array has not changed size. If I perform any operations on this array, it will ignore the elements where the mask is true.

```
numbers_array.min(), numbers_array.max()
(4, 42)
```

Sometimes you may want to filter by two conditions. For example, you may only want values greater than 8 but less than 30. In Python, I can combine two conditions using an and (&) statement.

```
masked_nums_and = (numbers_array > 8) & (numbers_array < 30)
masked_nums_and
```

```
array([False, False, True, True, True, False], dtype=bool)
```

So, the last value of the array (42) is removed now because of the second condition. Alternatively, you could use an or statement (|) to keep the numbers that are greater than 8 or less than 30. This will return the opposite mask from the previous example.

```
masked_nums_or = (numbers_array <= 8) | (numbers_array >= 30)
masked_nums_or
```

```
array([ True, True, False, False, False, True])
```

In the examples above, I showed only one or two conditions, but there is no limit to the number of conditions that you can chain, or combine, together.

4.6. Time Series Data

In the Earth sciences, scientists are quite often interested in time series data. Fortunately, handling time series data is straightforward with a little bit of practice. The datetime format is not built in to Python, but can be accessed from the **datetime** package:

```
from datetime import datetime, timedelta
```

Let's say I want to input the date September 22, 2004, 4:16 UTC. Datetime object attributes include year, month, day, hour, second, and microsecond (in that order). If any of the keywords are missing, they're set to the first instance of that parameter (e.g. month will be set to January, day will be set to 1, and so on). So not every attribute must be set. For instance, I did not set the microsecond attribute below. Additionally, time zone information can be entered, but it is a good idea to follow the scientific convention of working in UTC.

```
#September 22, 2004 at 4:16 P.M
flight_time = datetime(2004, 9, 22, 4, 16)
flight_time.isoformat()
```

```
'2004-09-22T04:16:00'
```

Then to view this in a more human-readable format, I can either use the `isoformat` method (formatting the date in YYYY-MM-DD HH:mm:SS) or use a custom format created using the `strftime` method.

```
flight_time.isoformat(), flight_time.strftime('%Y/%m/%d %H:%m')

('2004-09-22T04:16:00', '2004/09/22 04:09')
```

Converting to the datetime format is useful because you change dates without keeping track of the numbers of days, hours, and seconds that have elapsed:

```
flight_time_shift = flight_time+timedelta(days=101)
flight_time_shift.strftime('%h %d %Y')

'Jan 01 2005'
```

4.7. Loops

Loops help with repetitive tasks, whether it's opening many files sequentially or reapplying a function. In Python, there are two kinds of loops: `for` and `while`. **For loops** iterate over a sequence of numbers or a list of objects. The following code contains a `for` loop that prints each of the names to the screen:

```
team_members = ['Sayid', 'Claire', 'Jack', 'Hurley', 'Sawyer',
  'Sun-Hwa', 'Kate', 'Charlie', 'Locke', 'Desmond', 'Juliet']

for member in team_members:
  print(member)
Sayid
Claire
Jack
Hurley
Sawyer
Sun-Hwa
Kate
Charlie
Locke
Desmond
Juliet
```

From above, you can see that the general syntax for the commands is:

```
Header 1:
    Nested command or line of code under header 1
Header 2:
    Nested command or line of code under header 2
    ...
This code outside of header 1
```

Python uses indentation to indicate the beginning and the end of **code block**, which includes loops, modules, functions, and object definitions. Code blocks are multiple lines of code that are run as a unit. The spacing is required; Python will return an error or run inappropriately if the code is misaligned. When not using Jupyter Notebooks, it's better to use spaces (four spaces for each level of indentation by convention) and not tabs when formatting the code. Most text editors (Section 11.1.1) can be set up to insert spaces when the tab key is pressed.

`If` statements can be used with loops to filter the elements and execute other commands. For instance, in the example below, it instructs Python to only print the names that begin with J.

```
for member in team_members:
  if member[0] == 'J':
     print(member)
Jack
Juliet
```

From the example above, you will notice that print command is indented twice; this indicates that it is part of the `if` statement.

`For` loops can also be performed over an array of numbers using `np.arange`. On its own, `np.arange(0, 11)` will print a list of values from 0 to 10, which by default increment by 1, skipping the last value. Below I will append numbers to an empty list. Note that the print statement is not inside the loop because the indentation is aligned with the `for statement.exp_numbers=[]`

```
for i in np.arange(0, 11):
    exp_numbers.append(i**2)
print(exp_numbers)

[0, 1, 4, 9, 16, 25, 36, 49, 64, 81, 100]
```

A useful function in `for` loops is `enumerate`. When called, it will return a **tuple** with both the index `i` and the value `exp_number` of the element in the list. A tuple is like a list, but the order and values cannot be rearranged or changed.

```
for i, exp_number in enumerate(exp_numbers):
    print(i, exp_number)
0 0
1 1
2 4
3 9
4 16
5 25
6 36
7 49
8 64
9 81
10 100
```

For loops end when the list is complete. **While loops**, on the other hand, can carry on indefinitely until a condition is met. For instance, let's print out all numbers that are not evenly divisible by 5. I also introduce `continue` and `break`, which can help stop the code if some conditions are met.

```
div_by_5 = False
i = 0

while not div_by_5:
    num = exp_numbers[i]

    if num == 0:
        i=i+1
        continue

    if (num%5 == 0):
        print(num)
        div_by_5 = True
    else:
        i=i+1
        if i >= len(exp_numbers):
            div_by_5 = True
            break
25
```

In the loop above, the first `if` statement checks if the integer is zero. I do this because integer division of 0 yields a mod zero even though not technically a factor of 5. The integer *i* is first incremented, and the continue statement restarts the loop from the top.

The second if statement checks if the value is divisible by 5; if it is, they the number is printed and the flag is set to true, ending the while loop.

The break command tells Python to immediately stop executing the loop, even if the while condition loop hasn't been met. In the above example, this was done to prevent the code from executing past the end of the array and throw an error. Another reason to add break commands is to prevent infinite looping.

For loops are useful when you know how many times you want to execute the loop. For instance, if you have an array with 100 elements and you want to multiply each by 8, then a for loop is appropriate. There are also situations where you do not know in advance how many times you want the loop to run. For example, a while loop is appropriate if you want to iterate over an array until you find the first value greater than 8.

4.8. List Comprehensions

Loops are common in other programming languages. However, they are usually slow in Python. Python programming philosophy encourages programmers to keep code simple and readable, which the community calls **Pythonic**. The loop examples in the previous section illustrate how to use continue and break, but the code is not Pythonic: it is overly verbose for a simple task. **List comprehensions** are a computationally efficient and readable way to perform a repeating operation and return a new list variable. Because the list comprehension iterates on each element, they can be used in place of do and for loops for simple tasks. The syntax for list comprehensions is as follows:

```
new_list = [expression(item) for item in old_list if condition(item)]
```

In the pseudocode above, the expression is any operation, whether I want to just return the item value from old_list or perform a mathematical operation on it (expression). Item is the element of the list that I am iterating on. I can add an optional condition to filter the results. I can use list comprehensions on all the loops in the previous section. For instance, in the previous section, I wanted to print the list of team members that have names that begin with 'J':

```
[member for member in team_members if member[0] == 'J']
```

What was previously three lines of code is now one. More important than line count, the above statement is readable. The second loop from Section 4.7 takes the exponent of each item, but does not apply any filtering on the list.

```
exp_numbers_list = [value**2 for value in np.arange(0, 11)]
print(exp_numbers_list)
[0, 1, 4, 9, 16, 25, 36, 49, 64, 81, 100]
```

Note that the returned value above is a list and not a NumPy array, even though I am iterating over a NumPy array inside the list. However, the returned list can be quickly converted back to NumPy if necessary. Although the above is an improvement, the code below achieves the same result but even more quickly:

```
list(np.arange(0, 11)**2)
[0, 1, 4, 9, 16, 25, 36, 49, 64, 81, 100]
```

The last loop example (in Section 4.7) was not easy to read or understand. I am looking for the first value that is divisible by five. I need two conditions: one that the item is not 0 (`value > 0`) and another to check that there is no remainder when dividing by 5 (`value%5 == 0`). Since the result will be returned as a list of all items matching our criteria, I can select the first instance by subsetting the list using [`0`]:

```
filtered_values = [value for value in exp_numbers_list if (value%5 == 0)
  & (value > 0)]
print(filtered_values[0])
25
```

Although I illustrate different methods for certain tasks, my philosophy is to first understand the problem, make a first attempt at solving it, and then consider a more elegant way to solve it. Over time, you will refine your skills and your code will become more readable. Sometimes, however, coders fear sharing code because it is not yet "tidy" enough, or they become frustrated with how long the task of refinement is taking – I want to caution against fixating on rewriting your code to achieve perfection. Do not let perfect be the enemy of good. Chapter 12 provides further guidance for improving your code, and there are great resources written by expert software developers for you to draw inspiration from (Appendix C).

4.9. Functions

So far, I have worked with built-in or imported functions and methods. You can also define your own functions in Python. For instance, if you want to write a function that will print "Hello Earth," it would be written using the `def` keyword:

```
def hello_earth():
    # Prints a greeting to our planet
    print("Hello Earth")
```

The # is used to write block comments and document your code. Statements following the # to the end of the line are ignored by Python. If you run the above code block, no output will be return. The previous code block only **defined** the function. To instruct Python to execute the command, I would **call** or run it like I did other command in this section:

```
hello_earth()

"Hello Earth"
```

Note that this function does not require any input data (**arguments**) or **return** any values. For instance, if I want to make a Celsius to Fahrenheit converter:

```
def celsius_to_fahrenheit(Tc):
"""This function will take a temperature in Celsius and
convert to Fahrenheit.

Arguments:
    Tc: Temperature in Celsius

Returns:
    Tf: Temperature in Fahrenheit
"""
    Tf = Tc*(9/5)+32
    return Tf
```

Just like with loops, the indentation is mandatory for the code inside the function. The `Tc` argument is mandatory, which is then passed into the function and used to evaluate `Tf`. Docstrings (strings beginning and ending with three quotes, either """ or ''') are also useful for longer documentation. Note that comments do not need to be indented, since the interpreter ignores them. The second-to-last line is the equation to perform the conversion. The return statement is essential because it passes the argument back to the main program. If you try to call `Tf` outside the function, you will get an undefined function error. Only values that are explicitly returned and written to a variable are accessible.

To test our function above:

```
temp_f = celsius_to_fahrenheit(40)
print(temp_f)
104
```

The above example is a positional argument (which is unnamed and order is preserved), but I can also use keyword arguments (which are named and can be passed in any order. Rather than writing two functions, you could write one function to convert Fahrenheit to Celsius that will perform the conversion based on the keyword (`Tf=` or `Tc=`, respectively) entered. The following `if` statement detects if a 'C' is passed into the function. If 'F' is passed in, the `else if` (`elif` in Python) is triggered. If neither condition is met (`else`), the function will print an alert message and no conversion will be made. Instead, the function will return a -999 value. It might look like this:

```
def convert_temp(T=-999, convert_to='C'):
    '''This function will take a temperature in Celsius and
convert to Fahrenheit.

    Arguments:
        T : Temperature in Celsius or Farenheit
        convert_to: conversion to

    Returns:
        Tout: Temperature in Fahrenheit, Celsius, or -999
if nothing is calculated
    '''

    if convert_to == 'C':
        Tout = T*(9/5)+32
    elif convert_to == 'C':
        Tout = (T-32)*(5/9)
    else:
        print("Unrecognized conversion")
        Tout = -999
    return Tout
```

The above examples pass a known quantity of inputs into the function, which are `T=` and `convert_to=` in the `convert_temp` function. If you do not know in advance how many inputs you wish to pass into the function, you can pass in any number into the function using $*$.

```
def daily_precip(*args):
    '''This function calculates total rainfall of any
number of arguments

    Arguments:
        hourly rain totals
    Returns:
        Total precipitation
    '''
    result = 0
    for arg in args:
        result = result + arg
    return result
```

When I compute hourly precipitation:

```
daily_precip(12, 20, 33)
65
```

The above example illustrates one way you can pass arguments into a function. Let's look at a different way of passing arguments in same function. I argue that using a single keyword and passing a list would be the "cleanest" method, since lists can have elements added to them outside of the function call. In the function below (`daily_precip_list`), the inut `arg` is a list of any length:

```
def daily_precip_list(args):
    result = 0
    for arg in args:
        result = result + arg
    return result

hourly_precip = [0, 12, 20, 33]
daily_precip_list(hourly_precip)
65
```

These previous examples are simple, but science is full of complex methods and equations whose functions can help improve the readability of the program. In general, functions are useful for:

- Organizing your program
- Reusing code elsewhere in the program
- Extending code when making future improvements
- Simplifying the flow of the program by abstracting some of the program details

I pass arguments into functions in a way that is both clear and reusable. In this text, I will primarily use functions and methods from community packages; however, functions are essential for developing larger programs. As you hone your Python skills, there's no doubt that you will write many custom functions to carry out your work.

4.10. Dictionaries

Dictionaries are useful for storing values that need to be looked up. Dictionaries are denoted with curly brackets ({ }) and individual elements are stored as either strings, numbers, or any combination of the two.

```
ocean_units = {
  "temperature": "C",
  "velocity": "cm/s^2",
  "density": "kg/m^3",
  "pressure": "dbar",
  "salinity": "ppt" }
```

Dictionaries are not ordered like lists, so they are not accessed with a numeric index. Instead, they are accessed using the dictionary name:

```
ocean_units['pressure']
'dbar'
```

Dictionaries are useful for storing information inside the code while accessing it multiple times. Once created, entries can be added to the dictionary:

```
ocean_units.update({"salinity" : "ppt"})
```

If the entry already exists inside the dictionary, the above code would be updated with the new value. The `.item()` command allows us to view or iterate on all elements of the dictionary:

```
ocean_units.items()
dict_items([('temperature', 'C'), ('velocity', 'cm/s^2'),
('density', 'kg/m^3'), ('pressure', 'dbar'), ('salinity',
'ppt')])
```

Dictionaries are also useful for passing into functions, which are denoted as `**kwargs` in online documentation for functions. However, the variable can take on any name, so long as it has the leading double asterisks. The first * is mandatory as it denotes iteration (like in the *args example), while the second is a coding convention to distinguish passing a dictionary item that isn't keyed (e.g. `*args`, which passes an unstructured series of values). For example, the function below iterates on each item in the dictionary and prints a sentence.

```
def print_units(**kwargs):
    # Prints the variable and its units in a complete
sentence
    for key, value in kwargs.items():
        print("The units for " + key + " are " + value)

print_units(**ocean_units)
The units for temperature are C
```

```
The units for velocity are cm/s^2
The units for density are kg/m^3
The units for pressure are dbar
The units for salinity are ppt
```

Passing `**kwargs` is useful because you are not forced to know in advance which arguments will be passed into the function.

4.11. Summary

This chapter highlighted basic Python syntax so that you can take your first steps. Before I can create visualizations, I will have to import datasets. There are numerous formats, and most are multidimensional, making them a bit challenging to work with at first. However, in this chapter, you have learned the basics of data manipulation and combination. In the next chapter, I will discuss these formats and import real, publicly available satellite datasets.

References

Oliphant, T. E. (2015). Guide to NumPy. CreateSpace Independent Publishing Platform

Slatkin, B. (2015). Effective Python: 59 specific ways to write better Python. Upper Saddle River, NJ: Addison-Wesley.

5

IMPORTING STANDARD EARTH SCIENCE DATASETS

This chapter contains tutorials that use Python packages to import common data formats in Earth science, which includes CSV, netCDF4, HDF5, and GRIB2. For CSV files, the Pandas package enables users to quickly import the data and display it in a table format that resembles a spreadsheet. Self-describing formats, such as netCDF4 and HDF5, can be imported and inspected using the netCDF4 and h5py packages. Users can inspect the file header, which describes the data variables, their dimensions, and ancillary information such as quality flags and fill values within Python. This chapter will also cover newer packages that have been further optimized for Earth science datasets, such as xarray. Also covered is how to access data using OPeNDAP, which allows users to download the data within a Python session.

Data are science's lifeblood. Observations can inspire scientists to ask questions, test hypotheses, and after careful examination of data, lead to scientific discoveries. However, in practice scientists may instead spend a lot of their time writing code so they can read the data and organizing it into a format that is ready for analysis. There are ongoing efforts to promote analysis-ready data (ARD), which make satellite data more accessible and easier to examine so that scientists can spend more time on scientific inquiry and less time on coding (Dwyer et al., 2018).

Python has many options for importing data through several beginner-friendly packages. As discussed in Chapter 3, common formats include text files or scientific data formats like **netCDF** and **HDF**. Each of these files requires an

Earth Observation Using Python: A Practical Programming Guide, Special Publications 75, First Edition. Rebekah B. Esmaili.

DOI: 10.1002/9781119606925.ch5

algorithm (often called a **reader** or **decoder**) to interpret the contents and display them in a format that makes inspection, analysis, and visualization possible. This chapter will show us how to import data that is in a standard format using many of the skills you learned in Chapter 4.

In this chapter, you will build on the examples created using the **NumPy** (Section 4.4). To help us import and organize data, I will introduce the following packages:

- **Pandas**
- **netCDF4**
- **h5py**
- **pygrib**
- **xarray**

If you are working through these examples on your local computer, I recommend ensuring that these packages are installed (Appendix A.2).

I will show examples that import several datasets. The data are available online at: https://resmaili.github.io/Earth-Obs-Py. Specifically, the datasets are:

- **campfire-gases-2018-11.csv.** A csv file that contains a time series of trace gases during the California Camp Fire in 2018 Nov.
- **JRR-AOD_v1r1_npp_s201808091955538_e201808091957180_c201808092049460.nc.** A netCDF file that contains Aerosol Optical Depth (AOD) retrieved from a Suomi NPP overpass on 2020 9 Aug.
- **3B-HHR.MS.MRG.3IMERG.20170827-S120000-E122959.0720.V06B.HDF5.** An HDF5 file that contains a snapshot of global precipitation estimates on 2017 Aug 27.
- **NUCAPS-EDR_v2r0_npp_s201903031848390_e201903031849090_c201903031941100.nc.** A netCDF file that contains three-dimensional temperature and trace gas profiles.
- **gfsanl_3_20200501_0000_000.grb2.** A GRIB2 file that contains GFS analysis.
- **sst.mnmean.nc.** A netCDF file that contains the mean sea surface temperature (SST) for November 2018.

Except for `campfire-gases-2018-11.csv`, these files above are publicly available and downloadable. Thus, importing these files with Python will provide you with a realistic experience working with data. If you wish to work with other datasets, Appendix E is a primer on how to order and download other files.

5.1. Text

In Chapter 4, I introduced the NumPy package and how to perform array and matrix operations. NumPy can also read text files using the `loadtxt` or `genfromtext` functions. While these functions can efficiently read data, when importing strings `loadtxt` requires the user to provide the string length of each column. I find importing data to be a more pleasant experience with the `Pandas` `read_csv` function. The function name is somewhat a misnomer, as `read_csv` will read any delimited data using the `delim=` keyword argument. If you are searching for code online, you may come across examples using the **Pandas** `read_table`. However, this command is deprecated, so I encourage you to instead use `read_csv`. Over time, these older functions may cease to work, which would prevent your code from running in the future.

The example below use dataset that contains trace gas concentrations, `campfire-gases-2018-11.csv`, a file in the data folder. This file contains a time series of trace gases during a major wildfire in California that were produced from NUCAPS, a satellite sounding algorithm. At this point, I will focus the discussion on importing text data, so this is one of the few custom datasets that I created for this text. I will explain NUCAPS in greater detail in Chapter 7, where you will create maps. I recommend that you inspect all files before you import them to help understand the data structure. The first row of that dataset contains the file **header**, which assigned a name to each column that describes what each column of data is. The data elements are also comma separated and appears to have consistent formatted. First, I import Pandas and read in the file:

```
import pandas as pd
fname = 'data/campfire-gases-2018-11.csv'
trace_gases = pd.read_csv(fname)
```

Next, I will inspect the contents within the notebook using the `head` function, which will return the first five rows of the dataset:

```
trace_gases.head()
```

Latitude	Longitude	Time	H2O_MR_500mb	H2O_MR_850mb	CO_MR_500mb	O3_MR_500mb
39.69391	-120.49820	2018-11-01 10:39:44.183998108	0.001621	0.008410	84.18882	46.428890
39.99611	-121.75796	2018-11-01 10:39:44.383998871	0.001781	0.004849	77.59622	44.728848
39.248516	-120.72343	2018-11-01 10:39:52.183998108	0.001681	0.005230	80.73684	46.114296
39.54766	-121.97622	2018-11-01 10:39:52.383998871	0.001817	0.005295	77.39363	44.257423
38.80253	-120.94534	2018-11-01 10:40:00.184000015	0.001604	0.004505	82.68274	47.121635

In addition to facilitating data import, Pandas automatically stores data in structures called **DataFrames**. The two-dimensional (rows and columns) DataFrames resemble spreadsheets. In Pandas, the leftmost column is the row index and is not part of the `trace_gases` dataset. To access a single column inside a DataFrame, you can use the column name to print all values:

```
trace_gases['H2O_MR_500mb']
```

```
0        0.001621
1        0.001781
2        0.001681
3        0.001817
...
```

Another way to access the column of data is the `trace_gases.H2O_MR_500mb`. I prefer `trace_gases['H2O_MR_500mb']` because, for example, you can set `var_name="H2O_MR_500mb"` and then access the column using `trace_gases[var_name]`. This version is easier for creating loops (Section 4.7) and functions (Section 4.9).

You may only need a small part of a dataset when you import it. For example, the `trace_gases` variable contains more columns than I use in the examples in this chapter. Either indices or a lists of indices can be used to access specific elements of a list or array (Section 4.5). The fifth to the last column can be extracted using `[:5]`. This notation is helpful because I do not need to know the array size.

The code below takes the DataFrame column names (using the `columns` attribute), extracts the names of the last five elements, and converts this to a list.

```
drops = list(trace_gases.columns[5:])
print(drops)
```

```
['CO_MR_500mb', 'O3_MR_500mb', 'CH4_MR_500mb',
'N2O_MR_500mb', 'CO2_500mb']
```

Now that I have a list of the columns that I want to ignore, I can use the `drop` command to remove the columns and their associated data from a DataFrame. The `inplace=True` option permanently overwrites the existing `trace_gases` variable.

```
trace_gases.drop(columns=drops, inplace=True)
```

However, be aware that if you run this notebook cell a second time, `trace_gases` will refer to smaller DataFrame, not the original one

I imported from the dataset. If you want to access these columns later in your code, you should not use the inplace option and instead write it to a new variable (e.g., `trace_gases_small = trace_gases.drop(columns=drops)`).

Now inspect the contents to see if the variable has fewer columns:

```
Trace_gases.head()
```

Latitude	Longitude	Time	H2O_MR_500mb	H2O_MR_850mb
39.693913	-120.49820	2018-11-0110:39:44.183998108	0.001621	0.008410
39.996113	-121.75796	2018-11-0110:39:44.383998871	0.001781	0.004849
39.248516	-120.72343	2018-11-0110:39:52.183998108	0.001681	0.005230
39.547660	-121.97622	2018-11-0110:39:52.383998871	0.001817	0.005295
38.802532	-120.94534	2018-11-0110:40:00.184000015	0.001604	0.004505

As mentioned earlier, the data in the `campfire-gases-2018-11.csv` file are consistently formatted with comma separators. In practice, you will likely find data with uneven column lengths, multiple types of delimiters, missing values, or another inconsistency that may cause `read_csv` to fail or read improperly. This may require you to open the data file line by line.

Below is an example of how to use loops (Section 4.7) to read each row and column of data. While more tedious than the using `read_csv`, manually reading data allows for cell-by-cell control. After importing, the data can be written to lists or an empty NumPy array and converted to Pandas if you want to work with DataFrames.

```
with open('data/campfire-gases-2018-11.csv') as data:
       for row in data:
           print(row)
           column=''
           for character in row:
               if character != ',':
                  column=column+character
                 else:
                   print(column)
                   column = ''
           break
```

5.2. NetCDF

To begin, I need to first import the **netCDF4** package. There are other modules that can open netCDF4 files, such as xarray (Section 2.4), which has the netCDF4 package as a dependency. So, it is useful to first understand the netCDF4 package.

First, I will import the netCDF4 module:

```
from netCDF4 import Dataset
```

I will start by opening the following file:

```
JRR-AOD_v1r1_npp_s201808091955538_e201808091957180_c201808092049460.nc
```

As you can see, the satellite dataset names are quite long. However, the dataset name is encoded to give us some quick information on the contents. The prefix indicates the mission (JRR, for JPSS Risk Reduction), product (aerosol optical depth, or AOD), algorithm version and revision number (v1r1), and satellite source (npp for Suomi NPP). The remainder of the name shows the start (s), end (e), and creation (c) time, which are each followed by the year, month, day, hour, minute, and seconds (to one decimal place). So, I can learn several important features of the dataset without opening it.

Next, I will use the Dataset function to import the above dataset.

```
fname='data/JRR-
AOD_v1r1_npp_s201808091955538_e201808091957180_c201808092049460.nc'
file_id = Dataset(fname)
```

This step does not actually import the data, but rather assigns a file ID number and reads in the header information. If I print the contents of the `file_id` variable, I will get a long list of the global attributes, variables, dimensions, and much more.

```
print(file_id)
<class 'netCDF4._netCDF4.Dataset'>
root group (NETCDF4 data model, file format HDF5):
Conventions: CF-1.5
...
satellite_name: NPP
instrument_name: VIIRS
title: JPSS Risk Reduction Unique Aerosol Optical Depth
...
day_night_data_flag: day
ascend_descend_data_flag: 0
time_coverage_start: 2018-08-09T19:55:53Z
time_coverage_end: 2018-08-09T19:57:18Z
```

```
date_created: 2018-08-09T20:49:50Z
cdm_data_type: swath
geospatial_first_scanline_first_fov_lat:  42.36294
geospatial_first_scanline last_fov_lat:    37.010334
geospatial_last_scanline_first_fov_lat:    47.389923
...
dimensions(sizes): Columns(3200), Rows(768),
AbiAODnchn(11), LndLUTnchn(4)

variables(dimensions): float32 Latitude
(Rows,Columns), float32 Longitude(Rows,Columns), int32
StartRow(), int32 StartColumn(), float32 AOD550
(Rows,Columns), ...
```

The output above is worth inspecting. From the first bolded line, this file follows netCDF4 CF-1.5 conventions, which I discussed in Section 3.2.3. Some of the information learned from the file name is also present: this product is the JPSS Risk Reduction Unique Aerosol Optical Depth (`title`) Level 2 product (`processing_level`), and the data was collected from the Suomi NPP (`satellite_name`) VIIRS instrument (`instrument_name`). The start (`time_coverage_start`) and end times (`time_coverage_end`) metadata fields are consistent with the filename. I recommend that you read netCDF file header contents, especially the first time you are working with new data. Note that you can also use tools like Panoply (https://www.giss.nasa.gov/tools/panoply/) to inspect the contents of the netCDF file outside of Python.

In context of this book, the most useful part of this header is the variable list and their dimensions. In the header printout above, the variable names cluttered, so below I use the `keys()` command to print them out one by one:

```
file_id.variables.keys()
odict_keys(['Latitude', 'Longitude', 'StartRow',
'StartColumn', 'AOD550', 'AOD_channel', 'AngsExp1',
'AngsExp2', 'QCPath', 'AerMdl', 'FineMdlIdx',
'CoarseMdlIdx', 'FineModWgt', 'SfcRefl', 'SpaStddev',
'Residual', 'AOD550LndMdl', 'ResLndMdl', 'MeanAOD',
'HighQualityPct', 'RetrievalPct', 'QCRet', 'QCExtn',
'QCTest', 'QCInput', 'QCAll'])
```

In the variable list above, some of these variables are large array while others are single values. In the next example, I will use the AOD500 variable, which is the primary product in this file. AOD is a unitless measure of the extinction of solar radiation by particles suspended in the atmosphere. High values of AOD can indicate the presence of dust, smoke, or another air pollutant, while low values indicate a cleaner atmosphere.

Note that while metadata is helpful, the most thorough description of a dataset is typically found in a product's algorithm theoretical basis document (ATBD). ATBDs describe the retrieval algorithms, primary variables, and quality control measures. I recommend familiarizing yourself with the technical documentation if you are heavily using any satellite product.

I can extract AOD using the `.variable` class:

```
AOD_550 = file_id.variables['AOD550']
```

The array containing the data is now accessible to the notebook. Note that the type (using `type(AOD_550)`) of `AOD_550` has the type `netCDF variable`. Next, I convert it to a NumPy array and print the dimensions:

```
import numpy as np
AOD_550 = np.array(AOD_550)
AOD_550.shape
(768, 3200)
```

The code snippet above uses the NumPy array's `shape` attribute, which reveals the dimensionality of the array. In this case, `AOD_550` is a two-dimensional array. If you look back to the header, you can see under variables that the dimensions of the rows and columns are (768) and (3200), respectively, as the dimensions of `AOD_550` is *(Rows, Columns)*.

Now let's look at the data:

```
print(AOD_550)
array([[-999.99902344, -999.99902344, -999.99902344, …],
       [-999.99902344, -999.99902344, -999.99902344, ...],
       ...,
       [-999.99902344, -999.99902344, -999.99902344, ...,],
       [-999.99902344, -999.99902344, -999.99902344, ...]],
       dtype=float32)
```

You will immediately notice that there are a lot of –999.999 values. In practice, AOD falls between –0.5 and 5.0. So, –999.9 is not a reasonable number for AOD. Instead, –999.999 indicates that the data here are missing (Section 3.2.2).

Data may be missing because they are outside of the scan or over regions where the retrieval algorithm was unsuccessful. For example, the AOD algorithm cannot be retrieved over cloudy scenes or at night.

The missing values need to be excluded from mathematical operations. For instance, if I take a mean of this dataset, the resulting value is too low for the realistic AOD range of –0.5 to 5 range:

```
AOD_550.mean()
-289.09647
```

As I learned in Section 4.5, you can remove unwanted values using NumPy. While I already know the –999.9 is the missing value, this is not standard across all datasets. For instance, –9999.9 and –1 are both common missing values. So, I recommend extracting this value from the variable metadata directly, which I print below for AOD550:

```
print(file_id.variables['AOD550'])
<class 'netCDF4._netCDF4.Variable'>
float32 AOD550(Rows, Columns)
    long_name: Aerosol optical depth at 550 nm
    coordinates: Longitude Latitude
    units: 1
    _FillValue: -999.999
    valid_range: [-0.05  5.  ]
unlimited dimensions:
current shape = (768, 3200)
filling on
```

Some of the attributes above may be familiar while others may be new. For instance, earlier I told you that AOD ranges from –0.5 to 5.0, which is also printed in the `valid_range` attribute. To facilitate interpretation, Table 5.1 lists a small subset of possible attributes following netcdf4 conventions. Unidata provides a complete list and description of possible attributes:
https://www.unidata.ucar.edu/software/netcdf/docs/attribute_conventions.html

To filter the missing values, I need to extract the `_FillValue` attribute. Then, I will show two possible methods to remove the missing values. The first is more verbose but I have found it to work faster when handling large datasets. A limitation of the first method is that it does not preserve the original dimensions of the dataset. The second method uses masked arrays (Section 4.5), which preserves the original structure of the data at the expense of computational speed.

Table 5.1 Common Attributes Found a netCDF file Containing Remote Sensing Data

Attribute	Meaning
Units	The units of the values in the field
`long_name`	A long descriptive name. This could be used for labeling plots, for example.
`_FillValue`	The fill value is returned when reading values that were never written. Common values include 0, NaN, –999.9, –999.
`missing_value`	The number used to indicate "missing" values (e.g., where no observations were made). This is slightly different than the fill value, which is the default value, but they may be the same number.
`Valid_min, valid_max` OR `valid_range`	The minimum and maximum values in the dataset. Numbers outside of this range are set to missing.
`scale_factor`	Used to pack data. Data are multiplied by this number to convert from an integer to a real value or a larger integer. Used to pack data.
`add_offset`	Data are added to his number to shift values to larger integer (to save disk space) or to create a float value.
`coordinates`	An auxiliary coordinate variable as any netCDF variable that contains coordinate data (e.g., Latitude and Longitude)

5.2.1. Manually Creating a Mask Variable Using True and False Values

Below I extract the `_FillValue` and assign it to a variable called `missing`. I print `missing` to confirm it is –999.9.

```
missing = file_id.variables['AOD550']._FillValue
print(missing)
-999.999
```

If you recall from Section 4.5, using logical operators will return arrays of True and False values. Since the edges contain a lot of missing values, I am printing an interior part of the array where there is a mixture of missing and nonmissing values.

```
keep_rows = AOD_550 != missing
AOD_550[50:60, 100], keep_rows[50:60, 100]
(array([-9.99999023e+02, -9.99999023e+02,  5.40565491e-01,
         6.31573617e-01,  1.57470262e+00,  5.97368181e-01,
         6.23970449e-01,  6.71409011e-01,  6.59592867e-01,
        7.61033893e-01], dtype=float32),
```

```
array([False, False,  True,  True,  True,  True,  True,
True,  True,  True], dtype=bool))
```

In the code above, the keep_rows array has a value of False wherever the values are –999.9. Otherwise, keep_rows returns True.

The last step is to subset the data using the keep_rows array of True and False values. NumPy will remove variables whose index is False:

```
AOD_550_filtered = AOD_550[keep_rows]
AOD_550_filtered
array([ 0.06569075,  0.08305545,  0.11872342, ...,
0.14237013, 0.16145205,  0.14933026],  dtype=float32)
```

Finally, I re-compute the statistics using the filtered data. The result is within a physical range for AOD:

```
AOD_550_filtered.mean()
0.41903266
```

NumPy is dropping cells in the method above, so AOD_550_filtered is smaller than AOD_550. Below I use the .size attribute to see how many cells were removed. You can also use .shape.

```
AOD_550.size, AOD_550_filtered.size
(2457600, 1746384)
```

So indeed, the filter worked. It is a good idea to check array sizes when you are subsetting to make sure your code behaved as expect.

5.2.2. Using NumPy Masked Arrays to Filter Automatically

Some of the basic masked array functions were outlined in more detail in Chapter 4 when I discussed basic Python syntax. It may take some time and energy to familiarizing yourself with masked arrays. One benefit of masked arrays is that they preserve the dimensions and content of the original array. Figure 5.1 illustrates how the full dataset (left) is combined with the mask (left) to filter missing values. The result is that the gray values are excluded when a mask is used.

In the netCDF4 package, if you extract variables using the square brackets ([:,:]) the data will automatically be converted into a masked array if the _FillValue attribute is present. Note that a dataset must be CF compliant (https://cfconventions.org/) for this to work as expected:

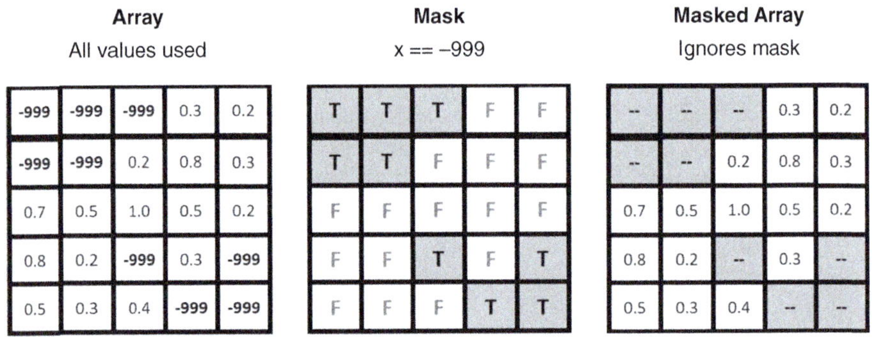

Figure 5.1 Illustration showing which values are used in a computation of a regular array and a masked array. White values ("False") are used in the calculation, while dark values ("True") are ignored. The original array shape is preserved.

```
AOD550 = file_id.variables['AOD550'][:,:]
type(AOD550)
numpy.ma.core.MaskedArray
```

Missing values will not be included in computations. Instead of showing -999.9, these values are dashed out.

```
AOD550[50:60, 100]
masked_array(data = [-- -- 0.5405654907226562
0.6315736174583435 1.5747026205062866 0.5973681807518005
0.6239704489707947 0.671409010887146 0.659592866897583
0.7610338926315308],

mask = [ True True False False False False False False
False False],
       fill_value = -999.999)
```

So, finally, I print out the mean AOD for this granule:

```
AOD550.mean()
0.41903259105672064
```

The mean value is the same using both methods. As you can see, the masked arrays method requires significantly fewer lines of code. As mentioned before, if you are working with large datasets, the code may run slower using masked arrays.

5.3. HDF

If you felt comfortable with the section on netCDF, you will find the steps for handling HDF files to be similar. I will import an HDF5 formatted file for global precipitation data from an algorithm called IMERG.

The file that I will open is:

```
3B-HHR.MS.MRG.3IMERG.20170827-S120000-E122959.0720.V06B.
HDF5
```

Like the Suomi NPP AOD example in the previous section, the file name can be decomposed to gather some information on the content. The filename includes the product name (*IMERG*), satellite (*MS*, for multi satellite), level (*3B*), algorithm version (*V06B*), and granule start time (*S120000*) and end time (*E122959*). The date is 20170827, which is when Hurricane Harvey was impacting Eastern Texas in the United States. The file extension for this file is `.hdf5`, but other common extensions include `.he5` and `.h5`.

I begin by importing the **h5py** package. You will see that the syntax is like that of the **netCDF4** library. The command to open the file and assign an ID is `File`. You may optionally want to add the `'r'` keyword to make it read only.

```
import h5py
fname = 'data/3B-HHR.MS.MRG.3IMERG.20170827-S120000-
E122959.0720.V06B.HDF5'
file_id = h5py.File(fname, 'r')
print(file_id)
<HDF5 file "3B-HHR.MS.MRG.3IMERG.20170827-S120000-
E122959.0720.V06B.HDF5" (mode r)>
```

Unlike netCDF, printing the `file_id` does not print the variables and dimensions in the file. Instead, you will have to use the `list` command:

```
list(file_id)
['Grid']
```

From the output above, you might think there is only one variable. This is not true; data in HDF files is often stored into groups. This dataset stored the variables within the *Grid* group, which acts as container for additional variables. In the example, the dataset only has one group (even though HDF files can have as many). There are can also be more groups within groups, depending on how the variables are organized.

If you `list` the items within the *Grid* group, you can see a full list of available variables:

```
list(file_id["Grid"].keys())
['nv', 'lonv', 'latv', 'time', 'lon', 'lat', 'time_bnds',
```

```
'lon_bnds', 'lat_bnds', 'precipitationCal',
'precipitationUncal', 'randomError', 'HQprecipitation',
'HQprecipSource', 'HQobservationTime', 'IRprecipitation',
'IRkalmanFilterWeight', 'probabilityLiquidPrecipitation',
'precipitationQualityIndex']
```

If you do not know in advance how many groups or variables are available, you can use the `visit` command with the print function passed to it to see all available variables and groups:

```
file_id.visit(print)
Grid
Grid/precipitationQualityIndex
Grid/IRkalmanFilterWeight
Grid/HQprecipSource
Grid/lon
Grid/precipitationCal
Grid/time
Grid/lat_bnds
Grid/precipitationUncal
Grid/lonv
Grid/nv
Grid/lat
Grid/latv
Grid/HQprecipitation
Grid/probabilityLiquidPrecipitation
Grid/HQobservationTime
Grid/randomError
Grid/time_bnds
Grid/IRprecipitation
Grid/lon_bnds
```

The two methods (using the `keys` or the `visit` command) are a useful way of inspecting the contents inside of Python.

Next, import the `precipitationCal` variable from IMERG, which combines all available microwave and infrared sensors to make a global precipitation dataset. Inspecting the header reveals that this is a three-dimensional variable (`time, lon, lat`), so it must be stored in a three-dimensional NumPy array called `precip`:

```
precip = file_id["Grid/precipitationCal"][:,:,:]
precip
array([[
       [-9999.90039062, ..., -9999.90039062],
       [-9999.90039062, ..., -9999.90039062],
       [-9999.90039062, ...,  -9999.90039062],
       ...,
       [-9999.90039062, ..., -9999.90039062],
       [-9999.90039062, ..., -9999.90039062],
       [-9999.90039062, ...,  -9999.90039062],
       ]],
       dtype=float32)
```

By looking at the `dtype` (types are discussed in Section 4.2), I see this is a float array. Unlike netCDF4, h5py does not automatically format the extracted data into a masked array. I will have to manually remove missing values ourselves. I also see a lot of -9999.9 values. I can examine the attributes to see if this is the `_FillValue`:

```
list(file_id["Grid/precipitationCal"].attrs)
 ['DimensionNames', 'Units', 'units', 'coordinates',
'_FillValue', 'CodeMissingValue', 'DIMENSION_LIST']
```

Conveniently, this attribute is called `_FillValue` just like in the netCDF4 AOD example in the previous section. This is a benefit of following common standards: even though I am examining a different dataset (different measurements, satellites, and publication entities), I know which attributes to inspect. Now I will assign the fill value to a variable:

```
missing = file_id["Grid/precipitationCal"].attrs
['_FillValue']
missing
-9999.9004
```

Next, let's make a masked array with the data and then compute the mean global precipitation.

```
 precip = np.ma.masked_array(precip, mask=PrecipMask)
 precip
masked_array(
data =
 [[[-- -- -- ..., -- -- --]
  [-- -- -- ..., -- -- --]
  [-- -- -- ..., -- -- --]
  ...,
  [-- -- -- ..., -- -- --]
  [-- -- -- ..., -- -- --]
  [-- -- -- ..., -- -- --]]],
 mask =
  [[[ True  True  True ...,  True  True  True]
   [ True  True  True ...,  True  True  True]
   [ True  True  True ...,  True  True  True]
   ...,
   [ True  True  True ...,  True  True  True]
   [ True  True  True ...,  True  True  True]
   [ True  True  True ...,  True  True  True]]],
 fill_value = 1e+20)
```

Finally, let's compute the mean:

```
precip.mean()
0.06620448132702939 9
```

Aside from the grouping, handling HDF and netCDF files in Python is almost the same. This is because netCDF files are a type of HDF file. One difference between the two file types is that netCDF files follow Climate and Forecast (CF) metadata conventions to ensure that datasets follow standard formats. For instance, CF conventions discourage variable grouping, so you will infrequently see operational netCDF with groups like I showed in the previous example.

5.4. GRIB2

The **pygrib** package (Unidata) has an interface between Python and the GRIB-API (ECMWF). ECMWF has since ended support for the GRIB-API as the primary GRIB2 encoded and decoder and now uses ecCodes. However, the package is still maintained by the developer (https://jswhit.github.io/pygrib/) and is useful for parsing NCEP weather forecast data.

```
import pygrib
```

The following code will import data from the Global Forecast System (GFS) GRIB2 file (stored in a variable called `gfs_grb2`) and create a list of the product definitions for each variable using a list comprehension (Section 4.8). The product definition will help us understand what the data variable is, how the variable is stored in the file and spatially gridded, and the time of the measurement.

```
filename = 'data/gfsanl_3_20200501_0000_000.grb2'
gfs_grb2 = pygrib.open(filename)
records = [str(grb) for grb in gfs_grb2]
```

There are 522 individual data product definition in this file (from the `len (messages)` command), so first let's inspect the contents of one line to start:

```
records[12]
13:Temperature:K (instant):regular_ll:isobaricInPa:level
40 Pa:fcst time 0 hrs:from 202005010000'
```

Colons (:) separate the sections of the product definition in this GRIB2 message. The elements are index (1), variable name and units (2–3), and spatial, vertical, and temporal definitions (4–8). There is one record for each pressure level and time. All variables can be extracted using the `.select(name=[variable])` command. Below, I select all the Temperature records (there are 46, which I see by using the `len(temps)` command). Since it is a long list, I am only printing some of these:

```
temps = gfs_grb2.select(name='Temperature')
print(temps)
...
301:Temperature:K (instant):regular_ll:isobaricInhPa:
level 80000 Pa:fcst time 0 hrs:from 202005010000,
315:Temperature:K (instant):regular_ll:isobaricInhPa:
level 85000 Pa:fcst time 0 hrs:from 202005010000,
 330:Temperature:K (instant):regular_ll:isobaricInhPa:
level 90000 Pa:fcst time 0 hrs:from 202005010000,
...
497:Temperature:K (instant):regular_ll:sigma:level 0.995
sigma value:fcst time 0 hrs:from 202005010000,
...
```

If you want to extract temperature at 85000 Pa, use the index (`315`) to pull that record:

```
temp = gfs_grb2[315]
```

Then, using `.values`, extract the data from the record:

```
temp.values
array([[249.30025, 249.30025, 249.30025, ... ],
[248.80025, 248.80025, 248.70024, ... ],
[249.20024, 249.20024, 249.10025, ... ],
...,
```

You can also extract the grid information and other import metadata for this record. To see all available information, use the `.keys()` command:

```
temp.keys()
... 'latitudes', 'longitudes', 'forecastTime', 'analDate' ...
```

The coordinates can be extracted using the `.latitude` and `.longitude`. The level, units, and forecast time can also be extracted from the file.

```
lat = temp.latitudes
lon = temp.longitudes
level = temp.level
units = temp.units
analysis_date = temp.analDate
fcst_time = temp.forecastTime
```

The above example illustrates the steps needed to open, inspect, and extract data from a GRIB file. However, there are some limitation to the approach. The data are multidimensional, so if you wish to extract multiple variables, pressure levels, and times, the resulting NumPy array will be quite large and complex. I will discuss xarray in the next section, which facilitates extraction and storage of complex datasets. Additionally, the cfgrib package can be called inside of xarray to open GRIB2 files, which I show in Section 5.5.3.

5.5. Importing Data Using Xarray

When I imported HDF, netCDF4 or GRIB2 data, I often converted these datasets into NumPy to mathematically operate on the data and Pandas to organize the data intuitively. Since nearly all remote sensing data are multidimensional, it is usually difficult to use Pandas and NumPy without subsetting the data. For instance, in the previous GRIB2 example, I extracted latitude, longitude, and variables like temperature into separate variables. **Xarray** (Section 2.4.1) extends the strengths of NumPy and Pandas to multi-dimensional data and is useful for working with large datasets. Additionally, the xarray can import a variety of common Earth science data formats. This is because xarray and Dask interface with Pandas, Scikit-Learn and NumPy to perform parallel processing and out-of-memory operations that can read data in chunks without ever being totally in the computers RAM. Note that xarray is a maturing package, so xarray's syntax may change with time and its capabilities are likely to increase. For new Python programmer, the learning curve for using xarray may be a bit steep. However, xarray syntax is very concise and most operations are optimized for speed.

5.5.1. netCDF

The xarray package can succinctly open **netCDF** files and structure variables into multidimensional arrays, called **DataArrays**, and organize them in xarray **Datasets,** which are intuitively similar to Pandas DataFrames. In this section, I will explore how to import netCDF data using xarray. Let's first import xarray and then open a NUCAPS datafile:

```
import xarray as xr
```

The `open_dataset` command will open the netCDF file and give you access to all the variables it contains. Below I import a NUCAPS dataset and print the header information, which includes the dimensions, coordinates, data variables, and attributes. A portion is reproduced below:

```
fname = 'data/NUCAPS-EDR_v2r0_npp_s201903031848390_
e201903031849090_c201903031941100.nc'
nucaps = xr.open_dataset(fname, decode_times=False)
print(nucaps)
<xarray.Dataset>
Dimensions: (Number_of_Cloud_Emis_Hing_Pts: 100,
Number_of_Cloud_Layers: 8,
Number_of_CrIS_FORs: 120,
Number_of_Ispares: 129,
Number_of_MW_Spectral_Pts: 16,
Number_of_P_Levels: 100,
Number_of_Rspares: 262,
Number_of_Stability_Parameters: 16,
Number_of_Surf_Emis_Hinge_Pts: 100)

Coordinates:
    Time (Number_of_CrIS_FORs) float64 ...
    Latitude (Number_of_CrIS_FORs) float32 ...
    Longitude (Number_of_CrIS_FORs) float32 ...
    Pressure (Number_of_CrIS_FORs,
```

```
        Number_of_P_Levels) float32 ...
    Effective_Pressure (Number_of_CrIS_FORs,
        Number_of_P_Levels) float32 ...

Dimensions without coordinates:
Number_of_Cloud_Emis_Hing_Pts,
Number_of_Cloud_Layers, Number_of_CrIS_FORs,
Number_of_Ispares, Number_of_MW_Spectral_Pts,
Number_of_P_Levels, Number_of_Rspares,
Number_of_Stability_Parameters,
Number_of_Surf_Emis_Hinge_Pts

Data variables:
    ...
    Temperature (Number_of_CrIS_FORs,
Number_of_P_Levels) float32 ...
    Cloud_Top_Fraction (Number_of_CrIS_FORs,
Number_of_Cloud_Layers) float32 ...
  ...
```

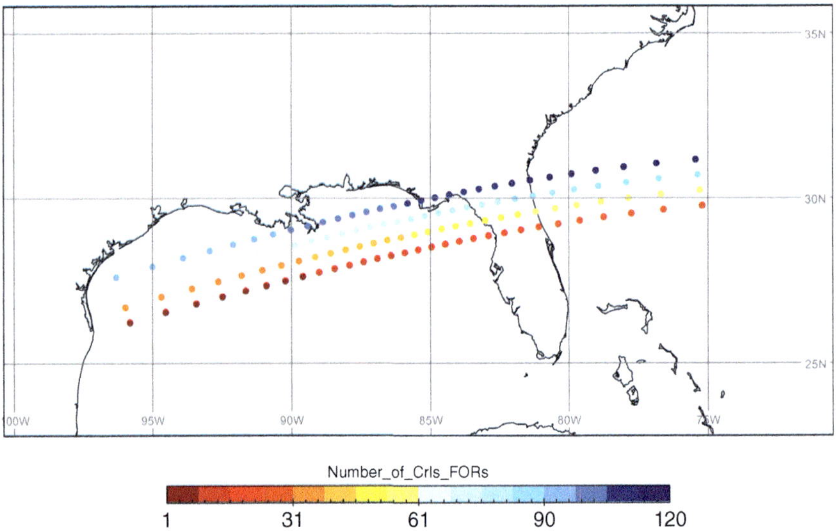

Figure 5.2 NUCAPS data are organized into 120 fields of regard (FORs) or footprints, each of which contain vertical profiles (100 fixed pressure grid points, from 1100 hPa to the top of the atmosphere). Each of the four scanlines (rows) has 30 footprints that are indexed by time, from left to right. In the afternoon orbit above, the earliest profile is in the bottom left (`index=0`) and the last is at the top right (`index=119`). Image is from 3 March 2019 at 18:48 UTC.

In the printout from the `xarray.Dataset` object, the values of the dimensions and coordinates are stored inside the `DataArray` object. For instance, the `Temperature` data variable contains a two-dimensional array of float variables, with the dimensions `Number_of_CrIS_FORs` (satellite footprint) and `Number_of_P_Levels` (pressure levels) which have a length of 120 and 100, respectively (Figure 5.2). The available coordinates for the data are `Latitude`, `Longitude`, `Time`, and `Pressure`. Not all variables have the same coordinates; for instance `Cloud_Top_fraction` has dimensions in `Number_of_CrIS_FORs` and `Number_of_Cloud_Layers` (of which there are eight).

5.5.2. Examining Vertical Cross Sections

Using xarray, I can easily subset the data using the dimensions. For example, perhaps I want to look at the vertical data in the first footprint. I can use the `.sel` command to index the DataArray for a specific dimension and save the result to a variable called `profile`. There are 100 values in this dimension and 0 will be the first index.

```
profile = nucaps.sel(Number_of_CrIS_FORs=0)
print(profile)
```

```
Dimensions: (Number_of_Cloud_Emis_Hing_Pts: 100,
Number_of_Cloud_Layers: 8, …,
Number_of_P_Levels: 100,
Number_of_Rspares: 262,
Number_of_Stability_Parameters: 16,
Number_of_Surf_Emis_Hinge_Pts: 100)

Coordinates:
    Time                      float64 1.552e+12
    Latitude                  float32 26.2246
    Longitude                 float32 -95.82371
    Pressure (Number_of_P_Levels) float32 0.0161 ...
1100.0
    Effective_Pressure (Number_of_P_Levels) float32 0.009

Data variables:
    Temperature (Number_of_P_Levels) float32 212.58 ...
    Cloud_Top_Fraction (Number_of_Cloud_Layers) float32
```

Looking at `Temperature` again, it is no longer a two-dimensional array, but rather one-dimensional over pressure values. The `Latitude` and `Longitude` coordinates are reduced to a single value (26.2246 and -95.82371, respectively).

In fact, Number_of_CrIS_FORs is no longer in the dimension list. Pressure is unchanged, since I did not subset this value. So, I have successfully reduced the dimensionality of the xarray dataset. There are many variables to work with, but if you want to access the data for just the temperature variable, use the following syntax:

```
temp_profile = profile.Temperature
print(temp_profile)
<xarray.DataArray 'Temperature' (Number_of_P_Levels: 100)>
array([212.5855  , 218.80322, 225.43922, 232.34381,
239.13559, 248.75534, 257.70367, 263.4195  , 265.9976  ,
266.10553, 262.76913, 258.74844, 254.02008, 248.83179,
244.12553, 240.88266, 238.28516, 236.50325, 235.0083  ,
233.87831, 232.78593, 231.43639, 229.92056, 228.23976,
      ...
      282.08627, 284.01132, 286.0352  , 287.5321  ,
289.10104, 290.48187, 291.65997, 292.64905, 293.26926,
293.85794, 294.6399  , 295.41217, 297.64713, 297.64713,
297.64713, 297.64713], dtype=float32)
Coordinates:
    Time                     float64 1.552e+12
    Latitude                 float32 26.2246
    Longitude                float32 -95.82371
...
```

Similarly, access the latitude and longitude values using profile.Latitude and profile.Londitude, respectively. However, as you can see from the printout above, Latitude and Longitude are single values; the item() command is helpful or extracting the number:

```
lat, lon = profile.Latitude.item(), profile.Longitude.
item()
print(lat, lon)
26.224599838256836 -95.8237075805664
```

5.5.3. Examining Horizontal Cross Sections

Perhaps next I want to look at the horizontal gradients, such as the spatial variation of temperature at 300 hPa. Instead of indexing over the footprint, I will index over the pressure level. For instance, I can create a mask to see where the pressure values are 300. I use the round() command on the Pressure DataArray because the values extend to four decimal places. Except for values near the top of the atmosphere, the levels are spaced greater than 1 hPa apart, so I can round to the nearest whole number.

```
mask = profile.Pressure.round() == 300
gradient = nucaps.sel(Number_of_P_Levels = mask)
print(gradient)
```

```
<xarray.Dataset>
Dimensions: (... Number_of_Cloud_Layers: 8,
    Number_of_CrIS_FORs: 120, ... Number_of_P_Levels: 1
    ...)
Data variables:
    ...
    Cloud_Top_Fraction (Number_of_CrIS_FORs,
    Number_of_Cloud_Layers)
    Temperature            (Number_of_CrIS_FORs,
    Number_of_P_Levels)
    ...
```

After subsetting the data horizontally, `Temperature` is still a function of `Number_of_CrIS_FORs`, `Number_of_P_Levels`. However, if you look at the dimensions at the top, the `Number_of_P_Level` size is 1 and not the original 100 levels.

In NUCAPS, the latitude and longitude positions repeat every 16 days, so I will extract the geolocation every time I open the file. However, the pressure levels are always the same in NUCAPS. So, rather than extract pressure from every footprint, I can create a dictionary (Section 4.10) to look up the index of the pressure level:

```
pres_dict = {}
for i, p in enumerate(profile.Pressure):
    if p >= 100:
        pres_dict.update({int(p) : i})
```

In the code above, I constructed a loop that iterates over each value inside the `Pressure` index of the footprint. The pressure levels are the same for every footprint, so I only iterate over the one profile. I am converting pressure to an integer (`int(p)` on the last line above) to make it easier to recall the pressure levels. Dictionaries cannot have repeating values, so I will ignore the top of the atmosphere values (where pressure is below 100 mb), which may round below 0. Another option is to construct a dictionary of float values, although then I would have to be able to recall the pressure level to four decimal places. Using the dictionary that I constructed above (`pres_dict`), I can access the index of the 300 hPa pressure level, which is 62:

```
pres_dict[300]
62
```

Then, I can select the individual pressure level without knowing the index:

```
gradient = nucaps.sel(Number_of_P_Levels = pres_dict
[300])
```

Overall, xarray is an excellent tool for managing larger datasets, but becoming fluid with it can take additional time. Additionally, xarray enables you to quickly inspect and extract important elements of the dataset. However, xarray relies on the source datasets adhering to CF compliance or at least partially so. If you are working with irregularly formatted netCDF files, some additional steps may be required, or you may need to use the netcdf4 package to manually extract the elements you need from your data.

5.5.4. GRIB2 using Cfgrib

Cfgrib is a maturing package that uses the ecCodes to decode GRIB2 data. The cfgrib engine can thus be used to import data using xarray, thereby taking advantage of capabilities for handling complex, multidimensional datasets. In the pygrib example from Section 5.4, I extracted air temperature from a single level, but it would be useful to extract values at all pressure levels or across multiple time steps. For data of European origin, I can use the `open_data` command like I did when I opened a netCDF file, with the addition of the `engine='cfgrib'` keyword argument:

```
gfs = xr.open_dataset(<filename>, engine='cfgrib')
```

Working with nonstandard GRIB2 files requires a few additional steps. In the GFS and NAM datasets, some variables have multiple pressure dimensions. For instance, air temperature records can have multiple vertical coordinates (which are `'isobaricInPa,'` `'surface,'` `'sigma'`, `'maxWind'`, `'heightAboveSea'`, and `'heightAboveGround'`), whereas data in European standard formats do not commonly do this. For the data to be stored in xarray in terms of latitude, longitude, and pressure, I need to ensure that all the coordinates are the same; otherwise, warnings or errors will be returned. Xarray has no way of distinguishing the variables in GRIB2 files because all air temperature measurements have the same name (`'Temperature'`). In a netCDF file, these fields may have different variable names (e.g., `'Temperature'`, `'Surface Temperature'`, `'Max Wind Temperature,'` etc.) and are therefore easier to parse because the dimensions must be consistent by design. I can use the `backend_kwards` command to specifically access the variables by vertical level type (`'typeOfLevel'`) and name

('name') using the filter_by_keys argument. Note that you must pass the filter_by_keys option as a dictionary object (Section 4.10). So, this will take the form of:

```
filter_keys = {'filter_by_keys' : {'typeOfLevel':
'isobaricInhPa', 'name': 'Temperature'}}
```

You can include as many or as few arguments here as you like to subset the GRIB2 file. Below I am accessing the air temperature on isobaric coordinates ('isobaricInhPa') using the 'typeOfLevel' key. If you re-examine some of the options printed in temps (defined in Section 5.4), you will see from index 419 that sigma is another available vertical coordinate grid. Thus, you could alternatively get the results in sigma coordinates from using the 'typeOfLevel': 'sigma' in the filter_keys dictionary above.

```
filename = 'data/gfsanl_3_20200501_0000_000.grb2'
gfs = xr.open_dataset(filename, engine='cfgrib',
backend_kwargs=filter_keys)
print(gfs)
<xarray.Dataset>
Dimensions: (isobaricInhPa: 33, latitude: 181, longitude:
360)
Coordinates:
    time              datetime64[ns] ...
    step              timedelta64[ns] ...
  * isobaricInhPa     (isobaricInhPa) int32 1000 975 950 …
  * latitude          (latitude) float64 90.0 89.0 88.0 87.0
...
  * longitude         (longitude) float64 0.0 1.0 2.0 3.0 ...
    valid_time        datetime64[ns] ...
Data variables:
    t                 (isobaricInhPa, latitude, longitude)
float32 ...
```

The temperature elements can now be accessed easily. I want all data at 850 hPa; I can subset it using .sel():

```
temp850hPa = gfs.sel(isobaricInhPa=850)
```

The 't' data variable (xarray is using the short name instead of the name field) can also be accessed. Since the data are global (with latitude and longitude coordinates), the values command prints a two-dimensional array of values.

```
temp850hPa.t.values
array([[249.30025, 249.30025, 249.30025, ..., 249.30025],
       [248.80025, 248.80025, 248.70024, ..., 248.80025],
       [249.20024, 249.20024, 249.10025, ..., 249.20024],
       ...,
       [240.00024, 240.00024, 240.10025, ..., 239.90024],
       [238.40024, 238.40024, 238.50024, ..., 238.40024],
      [238.40024, 238.40024, 238.40024, ..., 238.40024]],
dtype=float32)
```

Latitude and longitude can be accessed using `temp850hPa.latitude.values` and `temp850hPa.longitude.values`, respectively. To get the data for a specific point, further subset it:

```
temp850hPa_point = gfs.sel(isobaricInhPa=850,
latitude=90, longitude=0)
temp850hPa_point.t.values
array(249.30025, dtype=float32)
```

In this case, a single value is returned.

5.5.5. Accessing Datasets Using OpenDAP

Xarray supports the **Open-source Project for a Network Data Access Protocol** (OPeNDAP). OPeNDAP is a framework that allows software tools (including Python) to easily access a remote catalog of data. OPeNDAP can be found outside of Earth Sciences and allows users to access the data using a remote URL. It is stored in an "analysis ready" form. Access does require an internet connection (oPeNDAP, 2020).

In this example, I will open a mean sea surface temperature (SST) dataset. You can view available ESRL datasets on the website: https://www.esrl.noaa.gov/psd/thredds/catalog/Datasets/catalog.html. When you click on a netCDF4 filename, you will see a screen like Figure 5.3.

The catalog URL next to OPeNDAP shows you what needs to be opened using xarray. The syntax for importing the data is broken up into three lines so that it's more readable:

```
baseURL = 'http://www.esrl.noaa.gov'
catalogURL = '/psd/thredds/dodsC/Datasets/noaa.ersst.v5/
sst.mnmean.nc'
xr.open_dataset(baseURL+catalogURL)
```

Figure 5.3 Screenshot of the mean SST dataset in the NOAA/ESRL data catalog.

```
<xarray.Dataset>
Dimensions:     (lat: 89, lon: 180, nbnds: 2, time: 1995)
Coordinates:
  * lat         (lat) float32 88.0 86.0 84.0 82.0 80.0 78.0
...
  * lon         (lon) float32 0.0 2.0 4.0 6.0 8.0 10.0 12.0
...
  * time        (time) datetime64[ns] 1854-01-01 1854-02-01
...
Dimensions without coordinates: nbnds
Data variables:
    time_bnds   (time, nbnds) float64 ...
    sst         (time, lat, lon) float32 ...
...
```

An advantage of using OPeNDAP is that it does not permanently download the data to your machine. Rather, you store the data into memory directly and you can work with it inside your Python session or notebook. Additionally, you can bypass any data ordering and have immediate access to the datasets. However, accessing data through OPeNDAP does require a consistent, high-speed internet connection.

In addition to NOAA/ESRL, NASA's GES DAAC also has a list of available OpenDAP datasets. Note that a user account is required.

5.6. Summary

This chapter provided an overview of some of the syntax and tools needed to import data into Python. I worked with text, GRIB2, and two self-describing data formats, netCDF and HDF file formats. These are all common formats used in Earth remote sensing.

Before closing this chapter, I want to share a word of encouragement: downloading, importing, and preparing data can be one of the more frustrating aspects of scientific coding. Pre-processing data into a usable format takes time – sometimes more time than you have spent even thinking about the science question that you are trying to answer. Although the examples I showed in this chapter are fairly straightforward, in practice you will likely encounter more complex data. Do not be discouraged. Instead, take breaks, write out your problem, or walk through the code with a peer. It will get easier with time.

References

CF Conventions Home Page. (n.d.). Retrieved November 25, 2020, from https://cfconventions.org/

Dwyer, J. L., Roy, D. P., Sauer, B., Jenkerson, C. B., Zhang, H. K., & Lymburner, L. (2018). Analysis ready data: Enabling analysis of the Landsat archive. *Remote Sensing, 10*(9), 1363. https://doi.org/10.3390/rs10091363

ecCodes Home - ecCodes - ECMWF confluence wiki. (n.d.). Retrieved November 25, 2020, from https://confluence.ecmwf.int/display/ECC

NOAA/ESRL. Catalog https://psl.noaa.gov/thredds/catalog/Datasets/catalog.html. (n.d.). Retrieved November 25, 2020, from https://psl.noaa.gov/thredds/catalog/Datasets/catalog.html

SciTools. Iris examples — Iris 2.2.0 documentation. (n.d.). Retrieved April 15, 2020, from https://scitools.org.uk/iris/docs/latest/examples/index.html

The numpy.ma module — NumPy v1.15 manual. (n.d.). Retrieved April 15, 2020, from https://docs.scipy.org/doc/numpy-1.15.0/reference/maskedarray.generic.html

6

PLOTTING AND GRAPHS FOR ALL

Scientists need to communicate their results through plots. Tutorials in this chapter show the fundamentals of making plots using the Matplotlib package. Matplotlib is one of the oldest and well-supported visualizations packages within Python, and can support plotting data as histograms, bar plots, scatter plots, mesh plots, and contour plots. With a few lines of code, Matplotlib allows users to quickly customize aesthetics, such as fonts and color maps and combine data on a single plot or a panel of several plots.

One of the most important ways to communicate scientific research and environmental monitoring is through clear and readable plots. Python has several packages to create visuals for remote sensing data, either in the form of imagery or plots of relevant analysis. Of these, the most widely used packages is matplotlib. **Matplotlib** plots are highly customizable and have additional toolkits that can enhance functionality, such as creating maps using the **Cartopy** package, which I will describe more in Chapter 7.

In this chapter, you will use several packages that are helpful for importing, organizing, and formatting data: **NumPy, datetime, Pandas, xarray,** and **h5py**. The examples in this chapter use the following packages:

- **Pyplot, ticker,** and **colors** from matplotlib
- **Metpy**

If you are working through these examples on your local computer, I recommend ensuring that these packages are installed (Appendix A.2).

The examples in this chapter will visually examine the impact that the California Camp Fire, a deadly wildfire that began on 8 November 2018, had on local

Earth Observation Using Python: A Practical Programming Guide, Special Publications 75, First Edition. Rebekah B. Esmaili.

DOI: 10.1002/9781119606925.ch6

air quality. Figure 6.1 shows a true color imagery from VIIRS of the smoke plume. I will also use the following datasets:

- **VIIRSNDE_global2018312.v1.0.txt.** The VIIRS Active Fire Product can help us see how many fires were detected from space on 7 November 2018.
- **campfire-gases-2018-11.csv.** A time series derived from the NUCAPS satellite sounding retrieval to show the gradual increase in carbon monoxide (CO).
- **NUCAPS-EDR_v2r0_npp_s201903031848390_e201903031849090_c201903031941100.nc.**
- **MOP03JM-201811-L3V95.6.3.he5.** The November 2018 CO monthly mean from the Measurement of Pollution in the Troposphere (MOPITT), which is an instrument on the Terra satellite.

I will show examples of plots of increasing complexity to build awareness of matplotlib's capabilities. I will start with examples using one-dimensional data and progress to two- and three-dimensions and corresponding plot types and customization.

6.1. Univariate Plots

Plots with one variable (**univariate**) include frequency plots, which show counts of numerical ranges and categorical data. For example, to plot a distribution of

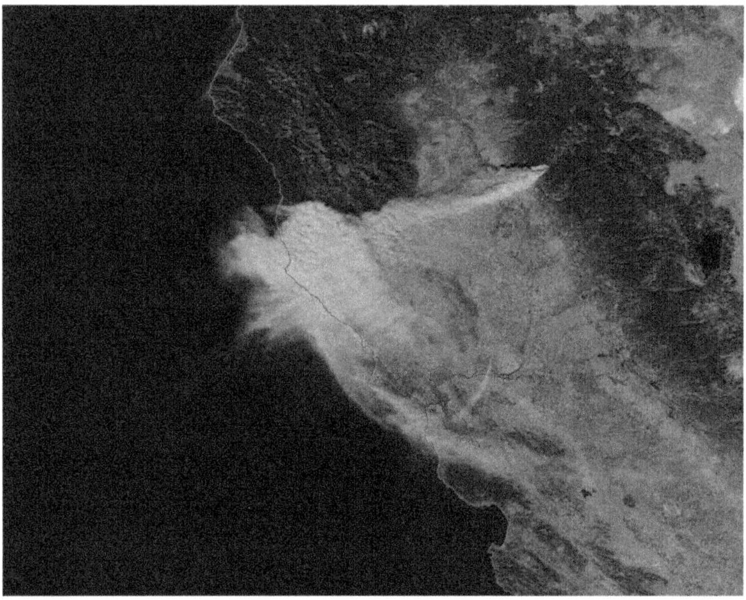

Figure 6.1 VIIRS imagery of wildfires on 8 November 2018. Source: Worldview (https://worldview.earthdata.nasa.gov).

values, the x-axis will show a number and the y-axis will show a total of how many times that number occurred in the dataset. However, if a dataset contains a long list of numbers that do not repeat in a precise way, counting the occurrence of each is probably not useful. **Binning** is a method of grouping continuous numbers into discrete intervals or discrete groups. For binned data, the x-axis will have a categorical name or bin range size while the y-axis shows population, count, or percentage.

6.1.1. Histograms

In this section, I will use the Pandas, NumPy, and Matplotlib packages, imported below:

```
import pandas as pd
import numpy as np
import matplotlib.pyplot as plt
```

In this example, Pandas is useful because `read_csv` will quickly load the text data from `VIIRSNDE_global2018312.v1.0.txt`, into a structured DataFrame. Recall from Section 3.2.2 that while csv stand for *comma separated variables*, the `read_csv` function in Pandas will import *any* separated text data using the `sep=` option, whether it be tab (`'/t'`), space (`' '`), a new line (`'/n'`) or other separation marker.

```
fires = pd.read_csv("data/VIIRSNDE_global2018312.v1.0.
txt")
fires.head()
```

	Lon	Lat	Mask	Conf	brt_t13(K)	frp(MW)	line	sample	Sat	YearDay	Hour
0	27.110006	30.769241		52	302.877533	5.814295	242	1735	NDE	2018312	1
1	26.083252	30.534357	9	100	332.959717	24.340988	301	1620	NDE	2018312	1
2	34.865997	28.162659	8	38	301.165985	6.107953	396	2589	NDE	2018312	1
3	34.872623	28.161121	8	71	307.277985	9.287819	396	2590	NDE	2018312	1
4	34.865070	28.158880	8	39	301.227783	6.001442	402	2590	NDE	2018312	1

In the DataFrame above, each row represents a single fire event detected for the year and day (the YearDay column) in the file. The fire intensity is determined by the brightness temperature (the `brt_t13(K)` column), and the fire radiative power is in megawatts (the `frp(MW)` column).

Even if you do not see any errors after importing a dataset, I recommend always opening the data file with a text editor or Excel to inspect the contents and make sure the imported data matches the source file.

Let's suppose I want to learn what the global distribution of fire radiative power is. From inspecting the `frp(MW)` column, these values extend to many decimal places. Rather than use a continuous scale, I can instead group the data into 10 MW bins, from 0 to 500 MW:

```
bins10MW = np.arange(0, 500, 10)
```

Now that I defined my bin size, I am ready to display my plot. Since Jupyter Notebooks are interactive, I can display all my plots inside my notebook. To do this, the entire contents of the plotting call must be inside the same code block. In the first line, `plt.figure` creates a blank canvas. Now, with each additional line, I am layering elements on this empty graphic. In the second line, I add the histogram to the figure using `plt.hist`, which automatically will count the number of rows with fire radiative power in the bins that I defined above in the `bins10W` variable. I must then pass the data (`fires['frp(MW)']`) and the bins (bins10MW) into `plt.hist`. The last command `plt.show()` tells matplotlib the plot is now complete and to render it:

```
plt.figure()
plt.hist(fires['frp(MW)'], bins=bins10MW)
plt.show()
```

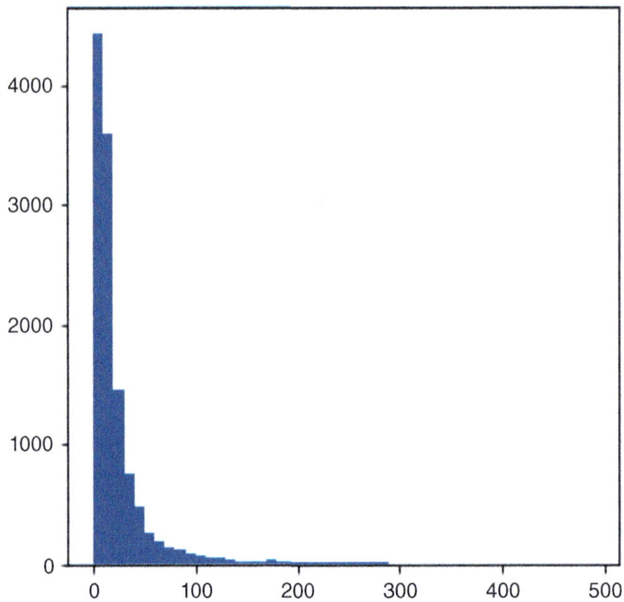

Now that I have the plot above, I want to render it again, but this time include better labels and scaling while keeping my code concise. For instance, I did not add any **aesthetics**, which are properties that make a plot more understandable, such as titles, labels, font sizes, annotations, or color schemes. In the next example, I will add labels to the x- and y-axis. Since there are thousands more fires with fire radiative power less than 100 MW than fires with higher values, the data are likely lognormal. The plot will be easier to interpret if I rescale the y-axis to a log scale while leaving the x-axis linear.

There are a few variations in how a plot can be called that can help simplify the code. The command `plt.subplot()` will return an axis object to a variable. In practice, common name choices include `ax` or `axes`. There are three numbers passed in, which correspond to rows, columns, and index. In the next few examples, I will use `plt.subplot(111)`, which is a subplot with one row and one column, and therefore, only has one index. In Section 6.3.6, I will show how to draw two figures either beside each other or stacked. The index defines which subplot is updated.

In the previous example, when I called `plt.hist`, matplotlib automatically updated the axis of the last used plot. While I only have one subplot in this next example, I recommend explicitly passing the plot axis in use. So, instead of `plt.hist`, I use `ax.hist`.

So, I build on the same code, but instead of `plt.figure`, I identify which axis (ax) using `plt.subplot` (see Table 6.1), then I will add labels using `set_xlabel` and `set_ylabel` lines 4–7:

```
plt.figure()
ax = plt.subplot(111)
ax.hist(fires['frp(MW)'], bins=bins10MW)
ax.set_yscale('log')
ax.set_xlabel("Fire Radiative Power (MW)")
ax.set_ylabel("Counts")
plt.show()
```

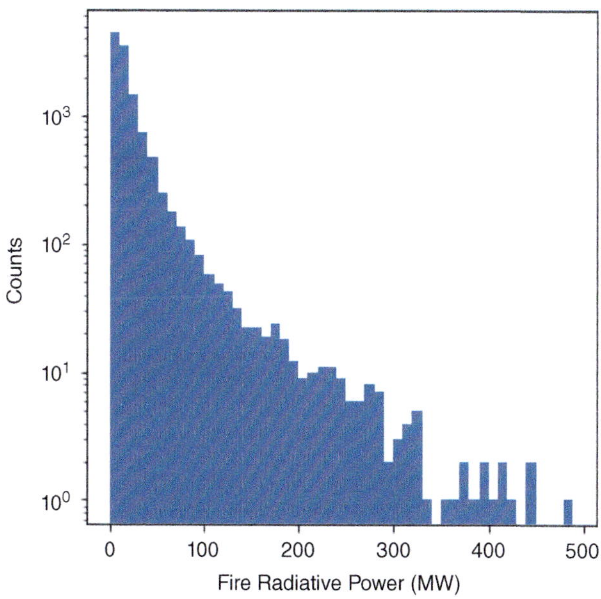

Table 6.1 Calls and Options for Initializing a Figure Using matplotlib

Figure call	Action	Returns	Options
`plt.figure()`	Creates a new figure	figures	`figsize, facecolor, edgecolor`
`plt.subplot()`	Adds a subplot to the current figure	axis	`nrows, ncols, index`
`plt.subplots()`	Combines the previous two calls Creates a figure and a grid of subplots with a single call	figure, axis, or an array of axes	`nrows, ncols`

In practice, I recommend starting with a simple plot when you are developing your visualizations. By adding incremental complexity, it is easier to debug your code if you have unexpected results. Because of the relative speed of developing plots in Python, it is typical to iterate on the same figure many times.

6.1.2. Barplots

In the previous example, I binned fire radiative into discrete numerical values. It is also possible to organize the data into categories. Below, I create a one-dimensional DataFrame to group the fires by their intensity into three arbitrary ranges with string names, which I define as low (0–100 MW), medium (100–200 MW), and high (200–1000 MW).

The `cut` command in Pandas splits the data into the three categories I defined. First, I store a list of the interval endpoints the `bins` variable. The labels variable is a list of the text names for each bin. Then, I call cut, pass in the one-dimensional data I want to bin, the bin ranges, and the labels. The `include_lowest=True` option specifies that I want to begin the intervals with 0–100, not 100–200. By default, the lowest bin value is excluded. I assign the output to the variable called `intensity` and print some of the results below:

```
ranges = [0.0, 100.0, 200.0, 1000.0]
labels = ['low', 'mid', 'high']
       intensity = pd.cut(fires['frp(MW)'], bins=bins,
include_lowest=True, labels=labels)
print(intensity)
```

```
0        low
1        low
2        low
3        low
...
11830    low
```

```
11831      low
11832      low
Name: frp(MW), Length: 11833, dtype: category
Categories (3, object): [low < mid < high]
```

The Pandas command `value_counts` will count the number of unique values. For the `intensity` variable, this action will count the number of rows with low, medium, and high. They will be stored in an indexed Pandas **series**, which is a one-dimensional array of any dimension, where our example has three rows:

```
intensity_counts = intensity.value_counts()
print(intensity_counts)
```

```
low      11418
mid        295
high       118
Name: frp(MW), dtype: int64
```

I will use the `ax.bar` instead of `ax.hist` because I already aggregated the data and have both one-dimensional x- and y-values to add to the plot. The low, mid, high categories are stored in the `index`, where I can access these values using `intensity_counts.index`. The y-values are stored in the `intensity_counts` variable from the last step. Like in the previous plot example, I am using the logscale and adding labels to the x- and y-axis below.

```
plt.figure()
ax = plt.subplot(111)
ax.bar(x=intensity_counts.index, height=intensity_counts)
ax.set_yscale('log')
ax.set_xlabel("Fire Radiative Power (MW)")
ax.set_ylabel("Counts")
plt.show()
```

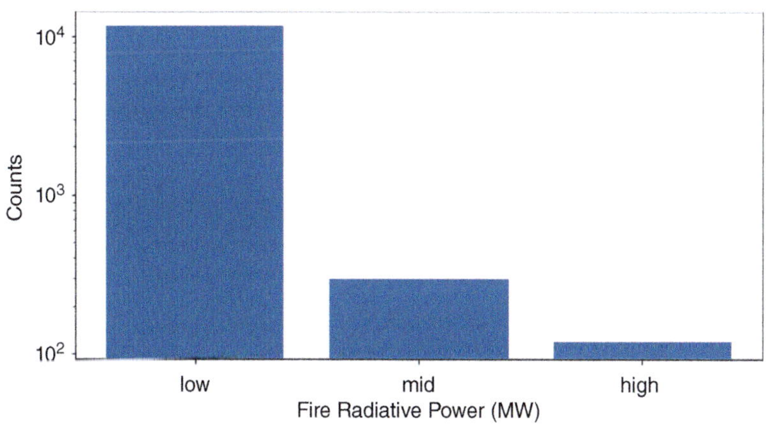

In the examples above, I used the same one-dimensional data for both histogram and bar plots. In the next section, I will show some examples of plots that utilize two variables to display smoke plume composition from satellite soundings.

6.2. Two Variable Plots

In the next example, I will inspect a timeseries of emitted gases such as carbon monoxide and ozone in addition to the environmental water vaper. I will use carbon monoxide and ozone data from the NUCAPS atmospheric sounding dataset. NUCAPS measures the vertical distribution of temperature, moisture, ozone concentration, and trace gases using infrared and microwave sounders. NUCAPS data are routinely stored as a granule in a netCDF file. To simplify the discussion in this section, I created a text file that contains a time series of NUCAPS trace gases within 150 km of main fire (39°N, 121°W). I will show an example using the original NUCAPS netCDF4 file in Section 6.2.7.

Using Pandas, import the `campfire-gases-2018-11.csv` dataset. You can inspect the column names either by manually opening the csv file or by printing it in your notebook, as I do below:

```
fname = 'data/campfire-gases-2018-11.csv'
trace_gases = pd.read_csv(fname)
trace_gases.columns
Index(['Latitude', 'Longitude', 'Time', 'H2O_MR_500mb',
       'H2O_MR_850mb','CO_MR_500mb',
       'CO_MR_850mb', 'O3_MR_500mb'
       'CH4_MR_500mb', 'CH4_MR_850mb', 'N2O_MR_500mb',
       'N2O_MR_850mb', 'CO2_500mb', 'CO2_850mb', 'Datetime'],
      dtype='object')
```

From the column names, the dataset contains the location, time, and the mixing ratios for water vapor, carbon monoxide, ozone, methane, nitrous oxide, and carbon monoxide. Water vapor is in mass mixing ratio, with units of kg of water to kg of dry air. All other trace gases are in volume mixing ratios, which are the volume of trace gas per volume of dry air. The volumetric mixing ratios are expressed in parts per billion (except for carbon monoxide, which is parts per million).

The pressure units are in millibar (mb), which is equivalent to hectopascals (hPa). NUCAPS has sensitivity to water vapor at 850 mb and 500 mb, which roughly corresponds to 1.5 km (5,000 ft) and 5.5 km (18,000 ft) above the Earth's surface for a standard atmosphere. The other trace gases are reported near 500 mb, where they have peak sensitivity. Details about the units are written in the variable metadata inside the source netCDF files.

The data in `campfire-gases-2018-11.csv` are a time series. It is useful to know how many days are in the file. In this example, the data are ordered sequentially, but not all datasets are. I recommend always using the `sort_values` function to explicitly define how the data are ordered. In the code example below, the `inplace=` option permanently overwrites the original `trace_gases` with the sorted values. Print the first (`[0]`) and last lines (`[-1]`) of `trace_gases` to see how many days of data are in the file:

```
trace_gases.sort_values(by='Time', inplace=True)
trace_gases.Time.values[0], trace_gases.Time.values[-1]
```

```
('2018-11-01 10:39:44.183998108', '2018-11-30
21:18:20.384000778')
```

The dates printed show that the file contains a month of measurements, spanning 1 November 2018 to 30 November 2018. These data can be used to examine how the trace concentration changes with time, before and after the fires begin.

6.2.1. Converting Data to a Time Series

Let's examine the contents of the datetime column below using the `head()` command, which prints the first five lines by default. In the printed output, the leftmost values are the index and the values on the right are the date and time value. At the bottom, Pandas also displays the column name (`Time`) and datatype (`dtype`), which is an `object`.

```
trace_gases['Time'].head()
```

```
0     2018-11-01 10:39:44.183998108
1     2018-11-01 10:39:44.383998871
2     2018-11-01 10:39:52.183998108
3     2018-11-01 10:39:52.383998871
4     2018-11-01 10:40:00.184000015
Name: Time, dtype: object
```

I can convert the `Time` column from an object format to a date (the `datetime` format, Section 4.6) to make the data easier to plot. I can tell Pandas how to interpret the string of text by using `formats`. After looking at the data above, the dates follow the pattern `%Y-%m-%d %H:%M:%S`, where `%Y` indicates a four-digit year, `%m` is a two-digit month, `%d` is a two-digit day, followed by two-digit hours (`%H`), a colon, and then minutes (`%M`), and then seconds (`%S`). The date portion of the time stamp is separated by dashes (-), the time by colons (:), with a space in between the two fields. There is no standard way of expressing dates in datasets, so if you encounter another format, you can refer to the entire assortment symbols at https://strftime.org/.

After looking at the data above, the dates are formatted as `%Y-%m-%d %H:%M:%S`, which I save as a string in the variable `fmt` in the code block below. The `to_datetime` function uses the format as a template to convert `trace_gases['Time']` from an object type to a datetime type.

```
fmt = '%Y-%m-%d %H:%M:%S'
trace_gases['Time'] = pd.to_datetime(trace_gases['Time'],
format=fmt)
trace_gases['Time'].head()
```

```
0    2018-11-01 10:39:44.183998108
1    2018-11-01 10:39:44.383998871
2    2018-11-01 10:39:52.183998108
3    2018-11-01 10:39:52.383998871
4    2018-11-01 10:40:00.184000015
Name: Datetime, dtype: datetime64[ns]
```

After converting the column format, `trace_gases['Time'].head()` shows that the `dtype` is now `datetime` and not an `object`.

6.2.2. Useful Plot Customizations

I want to introduce three new customization options that are vital for professional figures: limiting the *x*- and *y*-axis, using advanced text options in labels, and increasing the figure and font sizes in plots.

Matplotlib will automatically adjust the *x*- and *y*-axis limits (also called ranges) to fit the data if they are not explicitly stated in the code. However, it is often useful keep the *x* and *y* limits fixed, especially if you are comparing more than one plot. The `set_ylim` and `set_xlim` options pass custom minimum and maximum values to the axis object. In the next example, I will plot the 850 mb water vapor mixing ratio (`'H2O_MR_850mb'`), which can typically span 0 and 0.015. I want to display water vapor on the *y*-axis, so this interval is written as a list into a variable called `ylims`:

```
ylims = [0, 0.015]
```

Next, I want to show the time of the observation on the *x*-axis, and since the data are in the `datetime` format, the limits must also be in this format as well. Below, I import the `datetime` package, and using `datetime.date`, I can pass in 1 Nov. 2018 and 1 Dec. 2018:

```
import datetime
xlims = [datetime.date(2018, 11, 1), datetime.date(2018,
12, 1)]
```

Another useful customization is writing subscripts in the labels. Rather than labeling the *y*-axis "H2O," it is standard practice to write "H_2O." Matplotlib supports LaTeX typesetting commands, which makes it easier to use complex fonts in text and labels. Even if you are unfamiliar with LaTeX, thorough resources exist online, and I will show a simple example below. The syntax for using LaTeX in matplotlib labels is `$\mathregular{...}$`, where the {...} contains advanced typesetting options. The command `\mathregular` ensures that the subscript matches the default font used in your plots. Subscripts are produced using underscores (_) and superscripts using the caret (^). So, if I want to write H_2O in LaTeX, I will use `H_2O`:

```
ylabel = '$\mathregular{H_2O}$ mixing ratio'
```

Lastly, the default plot sizes and fonts are often too small for reading in presentations or publications. You can call `plt.rcParams.update()` to customize all the plots in your notebook with one line of code. For instance, the following command increases the figure size to 12" × 6" and the font size to 16 points for all future plots that are rendered in your notebook:

```
plt.rcParams.update({'font.size': 16, 'figure.figsize':
[12,6]})
```

The `figsize` can alternatively be passed into `plt.figure(figsize=(12,6))`, which will only change the sizes of the current plot, not all plots in the notebook. However, increasing the font size must be done one figure element at a time, so I recommend using the rcParams method in the previous code block.

6.2.3. Scatter Plots

A simple but useful way to visualize is using scatter plots (`plt.scatter`). Matplotlib requires scatter plot data to be array pairs of *x*, *y* coordinates. At the time of writing, you cannot easily pass in DataFrames into `plt.scatter`, but you can instead convert DataFrame columns into one-dimensional array using the `values` attribute:

```
mr = trace_gases['H2O_MR_850mb'].values
t = trace_gases['Time'].values
```

As I did for the histogram and barplots in Section 6.1, I initialize the figure and axis using `plt.figure` and `plt.subplot`. The new call below is `plt.scatter`, where I pass the *x*-axis data (`t`) and the *y*-axis data (`mr`), in that order.

In `plt.scatter`, the `c=` option allows us to enter a color (I chose `lightgray`). This is optional; if no color is specified, Matplotlib will use a default color scheme. If you wish to customize the colors, Matplotlib accepts some named colors and hexadecimal color values. Some great examples of color schemes are provided by Color Brewer 2.0 https://colorbrewer2.org/. I also encourage you to choose color-blind friendly palettes. Colors are discussed further in Section 12.4.3.

The next two lines define the x- and y-axis limits (`set_xlim` and `set_ylim`) and label to the y-axis. I could generate the plot now, but the x-axis text labels are long dates and overlap. `plt.xticks(rotation=35)` rotates the x-axis label 35° and makes the labels easier to read.

```
plt.figure()
ax = plt.subplot(111)
ax.scatter(t, mr, c='lightgray')
ax.set_xlim(xlims[0], xlims[1])
ax.set_ylim(ylims[0], ylims[1])
ax.set_ylabel(ylabel)
plt.xticks(rotation=35)
plt.show()
```

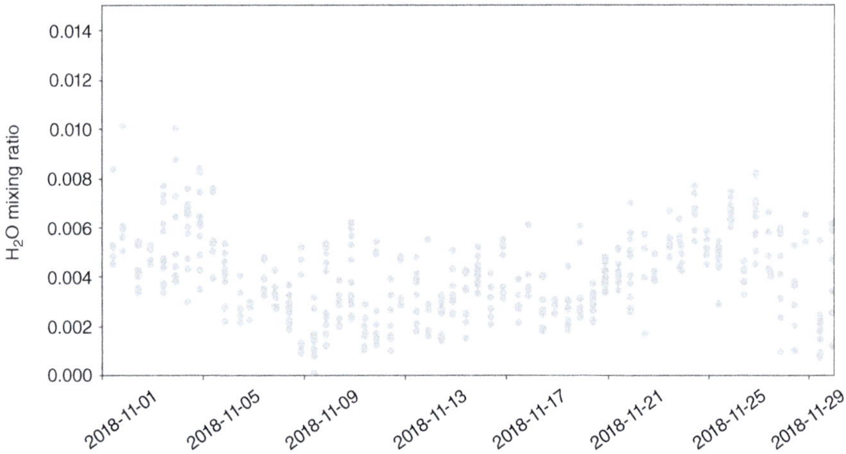

6.2.4. Line Plots

The scatterplot in the previous example shows that there are multiple observations per day. I can see the general pattern from the plot above; it can also be useful to add a line that shows the daily average as well.

One way to group or aggregate time series data is through the Pandas `resample` command, which takes two arguments: the desired level of aggregation and what column contains the `datetime` formatted data. In this case, I am aggregating by days (D). If you wanted to aggregate to three days, then you would use 3D, or Y for year, and so on. The on= option is set to the Time column. The resample command is only aggregating the data by day; I have to explicitly say what mathematical calculation to perform on the values. So, to compute the daily average, add the mean command to the end:

```
daily_average = trace_gases.resample('D', on='Time').mean()
```

Time	Latitude	Longitude	H2O_MR_500mb	H2O_MR_850mb
2018-11-01	39.461297	-121.227627	0.001126	0.006119
2018-11-02	39.520611	-121.171937	0.000464	0.004562
2018-11-03	39.589913	-121.175958	0.001372	0.005711

In the partial printing of daily_average above, you can see that there is an average value across all columns for each day for H2O_MR_850mb. Note on the far left that Time is now the index because I grouped by days. In the notebook, the word Time is vertically offset compared to the other column names. It is no longer a selectable column and if you try to print daily_average ['Time'], you will receive an error. To change it back to columns, you can call reset_index() to restore the Time column.

```
daily_average = daily_average.reset_index()
```

Alternatively, if you wish to keep Time as the index, you can also access the values by using daily_average.index. Retaining Time as the index could help speed up future mathematical operations in Pandas. Since I am not planning any further calculations, I will reset the index.

Now I can finally show the daily average mixing ratios of water vapor. The code below is similar to the previous example, except I now use plt.plot to add a line with the daily averages on top of the scatter plot. I use the linewidth= options to increase size (which is in pixels) and visibility on the plot. You may notice that at the top, I am returning the figure to the variable fig. In the previous example, matplotlib performed any operations using plt on the most recently called figure. By saving this object to a variable, I am clearly defining which figure the operations is performed on. Another change is that I removed plt.xticks(rotation=35) and replaced it with fig.autofmt_xdate(). This command will automatically rotate the dates so that they do not overlap.

```
fig = plt.figure()
ax = plt.subplot(111)

ax.scatter(t, mr, c='lightgray')
ax.plot(daily_average['Time'], daily_average
['H2O_MR_850mb'], linewidth=4)
```

```
ax.set_xlim(xlims[0], xlims[1])
ax.set_ylim(ylims[0], ylims[1])

ax.set_ylabel(ylabel)

fig.autofmt_xdate()

plt.show()
```

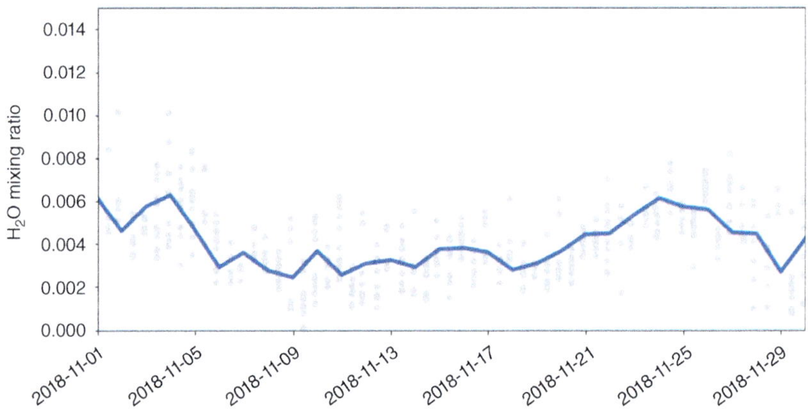

6.2.5. Adding Data to an Existing Plot

In practice, I often display more than one dataset on the same plot to compare the values. In the previous example, I created a plot that combined both a line and scatter plots. So, I can layer different datasets by calling `plt.plot` repeatedly. Below, I plot the 500 mb mixing ratio and 850 mb mixing ratio to show the moisture concentration at multiple heights in the atmosphere. Since they are the same plot type, Matplotlib will automatically pick a new color for each `plt.plot` command that is called on the same figure. Another way to distinguish the two lines is by adding a label option (`label=`) and a legend (`plt.legend`) to distinguish the two lines.

```
fig = plt.figure()
ax = plt.subplot(111)

ax.plot(daily_average['Time'],
daily_average['H2O_MR_500mb'], label='500mb')
```

```
ax.plot(daily_average['Time'],
daily_average['H2O_MR_850mb'], label='850mb')

ax.legend(loc='upper right')

ax.set_xlim(xlims[0], xlims[1])
ax.set_ylim(ylims[0], ylims[1])
ax.set_ylabel(ylabel)

fig.autofmt_xdate()
plt.show()
```

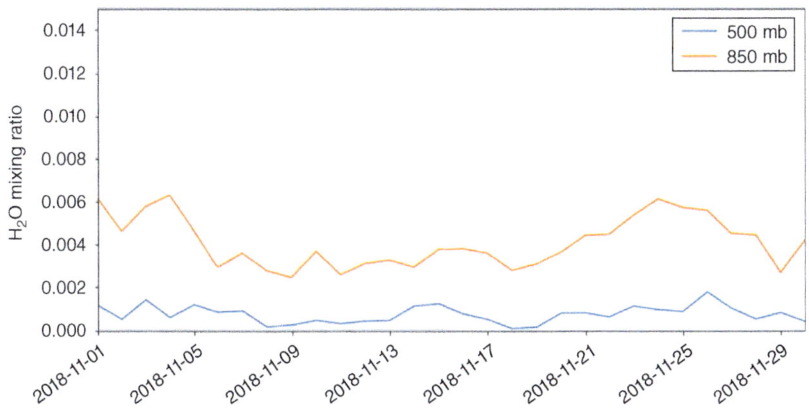

6.2.6. Plotting Two Side-by-Side Plots

In addition to overlaying two variables on the same plot, it is also useful to plot two or more figures side by side or in a grid. The arrangement of the plots is established in the `subplot` or `subplots` command, which have slightly different argument syntax. Table 6.1 summarized some of the differences, but in general:

- `plt.subplot` returns one axis for the given index and row/column number (the sample syntax is below). These options can be passed as either a continuous number (221) or (2, 2, 1). Confusingly, the index starts with 1 (unlike most other python indexing). Increasing the index starts in the upper-left corner and moves right across the row first, and then the column (Figure 6.2, black boxes). If you provide an index that is outside of the bounds of the columns and rows provided, you will get an error.

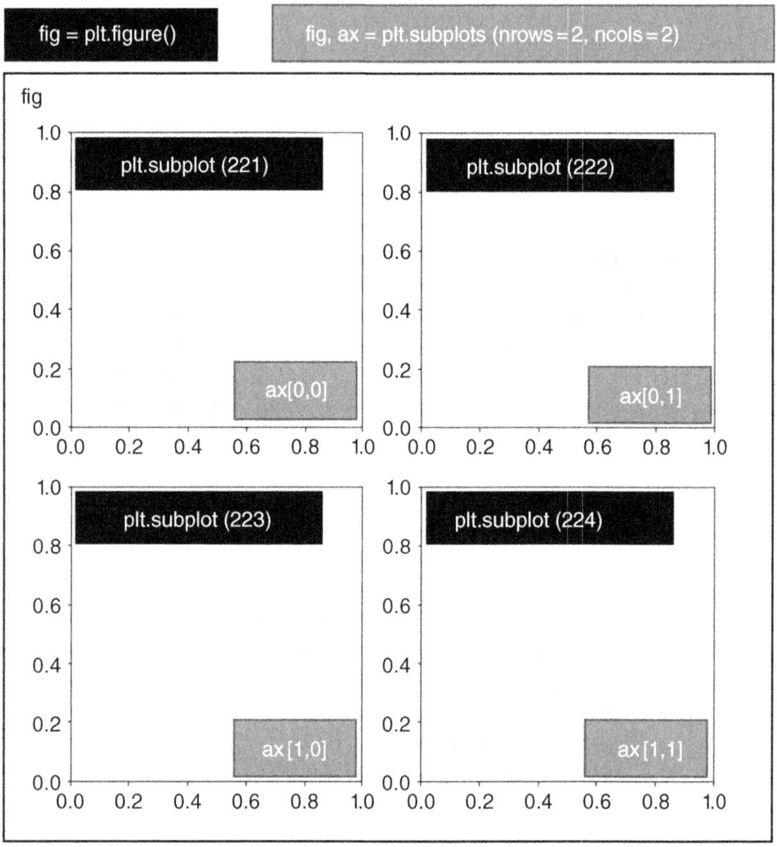

Figure 6.2 Different matplotlib axes layout syntax using the `plt.subplot` and `plt.subplots`.

```
fig = plt.figure()
ax = plt.subplot(row size, column size, index of active
plot)
```

- `plt.subplots` essentially combines `plt.figure` and `plt.subplot` and returns both figure and axis objects in a single command. However, the axis is returned as an array instead of a single value like in `plt.subplot`. You can then access the active axis using 0-based indices (Figure 6.2, gray boxes).

```
fig, ax = plt.subplots(ncols= row size, nrows=column
size)
```

You will find numerous examples of both formats online. I personally prefer `plt.subplots` because it is easier to see which axis is being updated.

Since I have the daily mean of all trace gases in daily_average, I will plot ozone and carbon monoxide. So below, you will notice two changes: the first line does not have plt.figure, but instead uses plt.subplots (with two columns). Second, instead of ax I specify ax[0] and ax[1]. The two gases have different limits, so would need to add separate set_ylim to fix the axis range. Since both are in units of ppm, I only put the *y*-label on the first plot. The *x*-axis covers the same span of days, though. I could use ax[0].set_xlim and ax[1].set_xlim, but for simplicity, I created a loop that iterates over all the ax and applies the limits.

```
fig, ax = plt.subplots(ncols=2)

ax[0].set_title('Ozone')
ax[0].plot(daily_average['Time'], daily_average
['O3_MR_500mb'])
ax[0].set_ylabel("Mixing Ratio (ppm)")

ax[1].set_title('Carbon Monoxide')
ax[1].plot(daily_average['Time'], daily_average
['CO_MR_500mb'])

for ax0 in ax:
    ax0.set_xlim(xlims[0], xlims[1])

fig.autofmt_xdate()
plt.show()
```

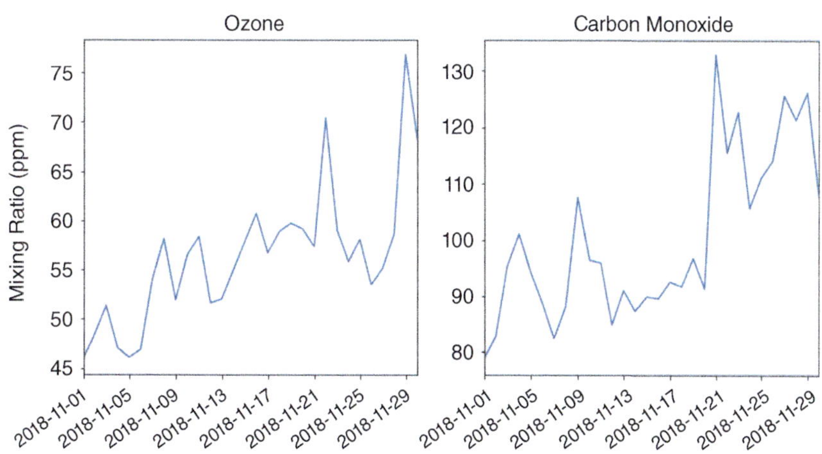

6.2.7. Skew-T Log-P

Radiosondes are instruments that measure temperature, pressure, moisture, and wind speed. Radiosondes are attached to weather balloons, which typically are launched twice a day (generally at 00 and 12 UTC) and take profiles up to 30 km (approximately to the 100 mb pressure level). Worldwide, radiosondes are released twice daily from 1,300 designated locations at 00 UTC and 12 UTC. These observations are useful for assimilation into forecasting models to improve their skill and to monitor air quality.

Satellite sounding products (such as NUCAPS) supplement surface observations by providing a larger spatial scale of soundings at times outside of the typical radiosonde launch. They are useful for assessing the preconvective environment ahead of hazardous conditions. There are some differences from surface observations, however. Radiosondes can take up to two hours to rise to the top of the atmosphere and drift as much as 200 km as they rise in the atmosphere. However, they are essentially a series of "point" observations. Satellite soundings, on the other hand, are instantaneous volume measurements. At nadir, the NUCAPS footprint is 50 km and 150 km at the scan edge; if there is significant variability on these scales, the retrieval will appear smoother than the radiosonde. So, while both can sample the vertical atmosphere, profiles from the two will not look identical.

Many forecasters prefer to view atmospheric profiles of temperature and moisture on Skew-T Log-P (often shortened to Skew-T) diagrams. Skew-T diagrams are the primary visualization used by meteorologists to express the thermodynamic state of the atmosphere. Unlike traditional two-dimensional profile plots, variables are plotted in log scale so that derived stability parameters can easily be calculated from looking at the plots. However, a challenge is that converting from water vapor to other units (such as dew point) does involve some complex unit conversions. For instance, the conversion to dew point from relative humidity is a function on temperature and pressure. Conveniently, the **MetPy** package can perform common conversions for you. MetPy is not included with Anaconda, you will need to download and install it (Appendix A).

To show a single profile, I will use xarray to import the NUCAPS netCDF4 file (Section 5.2.1). Below I import the packages that I will use:

```
import xarray as xr
from metpy.units import units
import metpy.calc as mpcalc
from metpy.plots import SkewT
```

Next, I import NUCAPS using `xr.open_dataset`.

```
fname = 'data/NUCAPS-EDR_v2r0_npp_s201903031848390_
e201903031849090_c2019030 31941100.nc'
nucaps = xr.open_dataset(fname, decode_times=False)
```

To make a Skew-T plot, I need to import the water vapor mixing ratio (`H2O_MR`), temperature (`Temperature`), and pressure (`Pressure`). Before I import these variables, I want to learn what the units (to convert to dew point temperature) and dimensions (to plot only one profile, not all 120 that are in the file) of each variable are. If I look at the attributes of the water vapor mixing ratio variable using `.attrs`, I can see that units are listed:

```
list(nucaps.H2O_MR.attrs)
```

```
['long_name', 'units', 'parameter_type', 'valid_range']
```

In fact, each variable inside the NUCAPS netCDF file has a `unit` attribute. Specifically, the units for water vapor, temperature, and pressure are:

```
nucaps.H2O_MR.units, nucaps.Temperature.units,
nucaps.Pressure.units
```

```
('kg/kg', 'Kelvin', 'mb')
```

Similarly, I can print the dimension names using `.dims`:

```
nucaps.H2O_MR.dims
```

```
('Number_of_CrIS_FORs', 'Number_of_P_Levels')
```

`Number_of_P_Levels` is an index of pressure levels whose values are stored in the `Pressure` variable. I will use all the pressure levels, so I do not need to subset this dimension. However, NUCAPS has 120 footprints or fields of regard (`Number_of_CrIS_FORs`). Each footprint contains a vertical profile with a diameter of 50 km at nadir and 150 km at the scan edge. To look at one profile, I will subset the array by selecting a single index (0–119) from `Number_of_CrIS_FORs` using the `sel` command from xarray. Below, I am arbitrarily choosing `Number_of_CrIS_FORs=1` and assigning it to a variable called `profile`:

```
profile = nucaps.sel(Number_of_CrIS_FORs=1)
```

The `sel` command returns the same contents as the original DataArray (including all the variables and attributes) but with fewer dimensions. Below, I can see that dimensions of `profile` now only include pressure when I print the `dims` of the variable `H2O_MR`:

```
profile.H2O_MR.dims
```

```
('Number_of_P_Levels',)
```

Similarly, temperature and pressure will also have fewer dimensions.

I need to import each variable to an array using `.values` to extract the data and `.flatten` to make it one-dimensional. The MetPy `units` function can facilitate making the Skew-T plot. I showed earlier that the temperature units are in Kelvin, the pressure units are in millibars (which is equivalent to hectopascals), and the water vapor mixing ratio units are in kilograms of water per kilogram of air. I can assign units to the array using `*units('K')` (for Kelvin) to any array:

```
T = profile.Temperature.values.flatten()*units('K')
MR = profile.H2O_MR.values.flatten()*units('kg/kg')
P = profile.Pressure.values.flatten()*units('millibar')
```

Using MetPy, I need to first convert mixing ratio to relative humidity (`relative_humidity_from_mixing_ratio`) and then relative humidity to dewpoint temperature (`relative_humidity_from_mixing_ratio`):

```
RH = mpcalc.relative_humidity_from_mixing_ratio(MR, T, P)
Td = mpcalc.dewpoint_from_relative_humidity(T, RH)
```

When you run the above code, you may see a **RuntimeWarning**, which is a message returned to the user after a potential problem is detected but the code continues to run. In the code block above, a warning was returned because there were negative log values. So, below I set the humidity values to 0 because these values are very high in the atmosphere and are very dry:

```
RH[RH < 0] = 0
```

Then, if I recalculate the dewpoint temperature, the errors will vanish:

```
Td = mpcalc.dewpoint_from_relative_humidity(T, RH)
```

Now that I have the pressure, temperature, and dewpoint temperature for out profile, I can construct the Skew-T plot. MetPy works in conjunction with matplotlib, so the code below will look like the other examples in this chapter. I return the figure object to the variable `fig`, which I pass into the MetPy function `SkewT` to create an empty Skew-T plot. Using the .plot command, I add the temperature (`T`) and dew point temperature (`Td`) to the plot. To help differentiate the lines, I use a dashed (--) line style on Td and add a legend to `fig`.

```
fig = plt.figure(figsize=[8,8])
skew = SkewT(fig)
```

```
skew.plot(P, T, label='Temperature')
skew.plot(P, Td, linestyle='--', label='Dewpoint')

fig.legend()
plt.show()
```

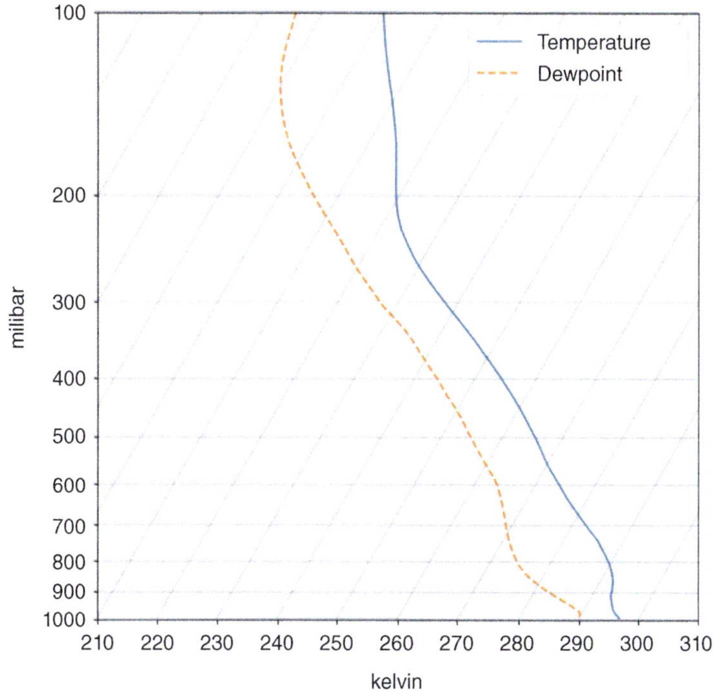

From the Skew-T, dewpoint temperature is much lower than temperature throughout the profile which indicates a dry environment. When combined with strong winds, this Skew-T shows that California was susceptible to wildfire prior to this event.

6.3. Three Variable Plots

Level 2 and Level 3 (Section 3.2.1) satellite data tends to be three-dimensional, with two positional arguments (latitude and longitude) and a physical variable (i.e. temperature, emissivity, soil moisture) becomes the third dimension. **Contour plots** and **mesh plots** are two useful ways of looking at three-dimensional data. I will again work the `fires` DataFrame in this section; if you have not done so, return to Section 6.2 and import the first two code blocks. Three-dimensional plots are some of the most important in remote sensing visualization, so I will continue discussing them in later chapters. Here, I will discuss some common situations where contour and meshplots are useful visualizations.

6.3.1. Filled Contour Plots

Contour plots are useful for finding local maxima in data and for displaying gradients in the data. Regular contour plots can be more easily distinguished using filled contours. Contours will artificially smooth data, which is usually not significant when looking at longer-term averages, such as those contained in Level 3 products or model data, which has high degrees of spatial correlation. These datasets are usually the easiest to plot in three dimensions.

Below, I am going to look at the monthly mean of the column total carbon monoxide from Measurement of Pollution in the Troposphere (MOPITT), which is an instrument on the Terra satellite. The data from November 2018 are in the file: `MOP03JM-201811-L3V95.6.3.he5`. The file extension (`.he5`) indicates that it is an HDF file, so I will import that data using the module **h5py** (Section 5.1.3).

```
import h5py
import numpy as np

fname = 'data/MOP03JM-201811-L3V95.6.3.he5'
file_id = h5py.File(fname, 'r')
```

Recall that to inspect the groups and variables, you can use the `visit` command:

```
file_id.visit(print)
```

This file has a lot of gridded values. I will import `RetrievedCOTotalColumnDay`, which is a two-dimensional variable in the `HDFEOS/GRIDS/MOP03/Data Fields/` group. I will also need latitude and longitude, which are both one-dimensional:

```
grp_name="HDFEOS/GRIDS/MOP03/Data Fields/"

co = file_id[grp_name+"RetrievedCOTotalColumnDay"][:,:]
lat = file_id[grp_name+"Latitude"][:]
lon = file_id[grp_name+"Longitude"][:]
```

The data above have a regularly spaced grid, so the latitude and longitude values need to be converted into a two-dimensional "mesh" in order to match the dimensions of CO. The `meshgrid` function will take the input one-dimensional arrays and copies the values into a two-dimensional matrix of dimensions [len(x), len(y)]. For instance, suppose I have two arrays:

If I compute the meshgrid, the returned matrices will have the three rows and two columns. The *x* term will repeat its values across the columns, and the *y* term will repeat values over rows.

Returning to the example, I will make a meshgrid of the one-dimensional latitude and longitude coordinates:

```
X_co, Y_co = np.meshgrid(lon, lat)
```

Before plotting, I need to check if all the dimensions match. However, after comparing the shape of co to X_co, I can see that the dimensions are flipped:

```
co.shape, X_co.shape
```

```
((360, 180), (180, 360))
```

To make the two arrays match, I can use the transpose() command to switch the *x*- and *y*-coordinates in co.

If you inspect the data, co has many -9999 values, which are likely missing values. In Section 5.2, I showed how to filter missing values using the fill value (which is usually stored in the _FillValue attribute of the variable). Below, I extract the fill value and save it to a value called missing. Then, I replace all missing values with np.nan so that they are not included in the plot:

```
missing = file_id[grp_name+"RetrievedCOTotalColumnDay"].
attrs['_FillValue']
co[co == missing] = np.nan
```

As I did in the proceeding sections, I call the plt.subplots() to generate the empty figure and axis. In the next line, I call ax.contourf and input the X_co, Y_co, and transposed co variables. co acts as a color value, which becomes the third dimension of the plot. I store this object into a variable co_plot so that I can pass it into ax.colorbar in order to map the colors to numeric values. I am omitting labels and further aesthetics to simplify the code:

```
fig, ax = plt.subplots()
co_plot = ax.contourf(X_co, Y_co, co.transpose())
fig.colorbar(co_plot, orientation='horizontal')
ax.grid(True)
plt.show()
```

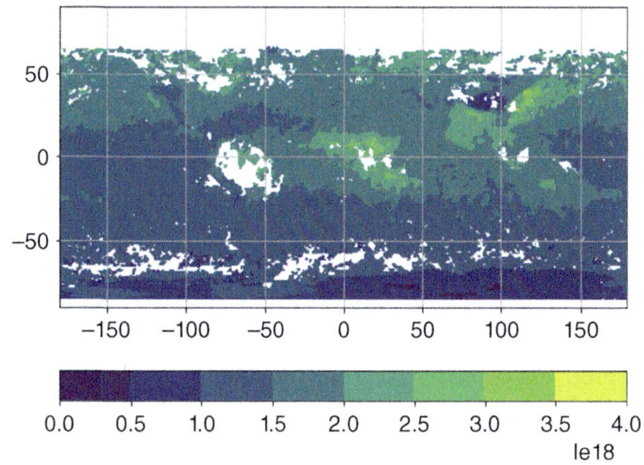

In the image above, you can see that there are regions where there are higher levels of CO (in molecules/cm^2). The data are clustered together and have global coverage, so a contour plot is a relevant choice in this scenario.

However, not all data are clustered together. For example, the `fires` is two-dimensional. In the univariate data section, I grouped data into categories based on the fire radiator power. If you recall, the `fires` also had the latitude and longitude coordinates of the fires. To make three-dimensional plots, you can look at the total fire counts with respect to latitude and longitude. I will use the round command on the latitude and longitude columns to round to the nearest 10 (the parenthesis indicates the number of decimal points, –1 thus equals 10). I do this because the coordinates are expressed to several decimal places, so I need to bin them in larger groups. I then assign the rounded latitude and longitude coordinates to the respective variables `lon_bin` and `lat_bin`.

```
fires["lon_bin"] = fires["Lon"].round(-1)
fires["lat_bin"] = fires["Lat"].round(-1)
```

Next, I use the `groupby` command from Pandas to count the total number of fires in each two-dimensional bin. This is similar to the `resample` command, but instead of using days I am using categories (`lon_bin` and `lat_bin`) to aggregate. Like resample, groupby does not perform any mathematical operation on the data. So, I use `size()` to count how many cells (and therefore fires) are in each bin:

```
fire_count = fires.groupby(["lon_bin", "lat_bin"]).size()
```

Again, like in resample, which moved the date to the index, the grouby will move `lon_bin` and `lat_bin` to the index. I use `reset_index` to shift `lon_` bin and lat_bin back into columns. Inside reset_index, I am also renaming the column containing the totals count; otherwise, Pandas will provide an undescriptive default column name.

```
fire_count = fire_count.reset_index(name="Count")
fire_count.head()
```

	lon_bin	lat_bin	Count
0	-160.0	60.0	3
1	-120.0	40.0	188
2	-120.0	50.0	142
3	-110.0	30.0	13
4	-110.0	40.0	4

The data are still in a **long form**, as in lon_bin, lat_bin, and Count are all one-dimensional arrays. All three must be reshaped to the same two-dimensional grid: lon_bin and lat_bin are the two-dimensional coordinates and Count will become the third dimension. The Pandas pivot command takes the DataFrame and reshapes it into a two-dimensional array structure using unique values of latitude and longitude. The column names become longitude (columns= 'lon_bin') where the row names become latitude (index='-lat_bin'). The counts is then assigned to the corresponding array cell value. If the new two-dimensional cell did not contain a value in count, it is filled with NaNs.

```
fire_count_3D = fire_count.pivot(index='lat_bin',
columns='lon_bin', values='Count')
fire_count_3D.head()
```

	lon_bin									
lat_bin	−160.0	−120.0	−110.0	−100.0	−90.0	−80.0	−70.0	−60.0	−50.0	−40.0
−40.0	NaN	NaN	NaN	NaN	NaN	NaN	3.0	NaN	NaN	NaN
−30.0	NaN	NaN	NaN	NaN	NaN	NaN	9.0	67.0	9.0	NaN
−20.0	NaN	NaN	NaN	NaN	NaN	NaN	4.0	42.0	2.0	1.0
−10.0	NaN	NaN	NaN	NaN	NaN	NaN	4.0	2.0	31.0	174.0
0.0	NaN	NaN	NaN	NaN	NaN	1.0	15.0	28.0	302.0	163.0

5 rows × 31 columns

Latitude and longitude can be transformed using NumPy's `meshgrid` command:

```
X_fires, Y_fires = np.meshgrid(fire_count_3D.columns,
fire_count_3D.index)
```

All three variables are now in the same dimensions, so I am ready to plot. I want to enhance the aesthetics in this next plot, so I am importing the ticker and colors packages from matplotlib.

```
from matplotlib import ticker, colors
```

As I did in the proceeding sections, I will call `ax.contourf` with `fire_count`. I use the `locator=ticker.LogLocator()` option because the fire frequency is lognormal (as shown in Section 6.2.1). This option will rescale the contours to log scale instead of plotting on a linear one. I overlay the contour with the original points (using `plt.scatter`) to compare with the original data. Optionally, `ax.grid (True)` creates the gridlines and is yet another optional aesthetic.

```
fig, ax = plt.subplots()

fire_plot = ax.contourf(X_fires, Y_fires, fire_count_3D,
locator=ticker.LogLocator())
fig.colorbar(fire_plot, orientation='horizontal',
shrink=0.5, extend='both')

ax.scatter(fires['Lon'], fires['Lat'], s=0.5, c='black',
alpha=0.1)

ax.set_xlabel('Longitude')
ax.set_ylabel('Latitude')

ax.grid(True)

plt.show()
```

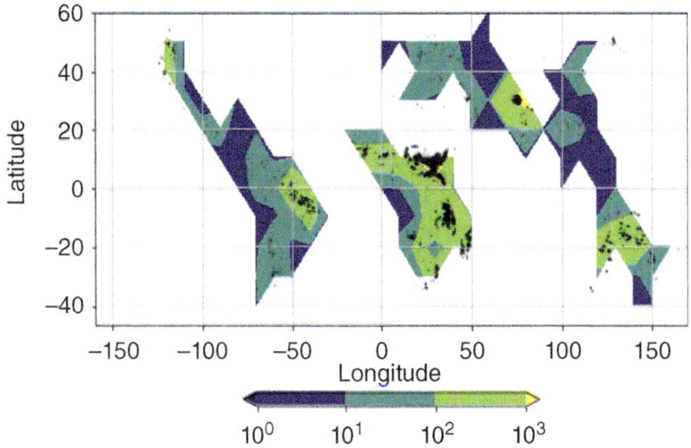

Unlike the first example, the above plot is not an effective visual. The contours are filling in areas that have few or no events because the data are unevenly distributed across the world. Instead, a `meshplot` would be more appropriate for this type of data.

If you briefly inspect the plot, you can start to see the outlines of continents. There are certain regions experiencing many fires (including California). In the next chapter, I will discuss how to project results such as these into maps.

Before I proceed to the next section, I go over some color bar customizations. In the previous example, I let matplotlib automatically place and scale the colorbar. This automation is helpful for a "quick look." However, it can be frustrating if you are creating a lot of figures, as the color bars often readjust the plot size. So, if you are comparing two different, the axes might not line up and the color scales will not be consistent across figures. Table 6.2 shows some of the more useful controls for the color bar. It is best to specify where the color bar is located using the `orientation` keyword, which can be placed vertically or horizontally. `Shrink` is also helpful, as it will shorten the bar. `Pad` will create more space between the color bar and the axes, so it is useful if the color bar covers up the labels.

Table 6.2 Helpful Options for matplotlob Colorbars

Property	Options	Description
orientation	vertical, horizontal	Specify if you want the colorbar below or to the right of the plot
fraction	0–1	Sets fraction of axes to use for colorbar
pad	0–1	Sets fraction of axes between colorbar and plot axes
shrink	0–1	Sets fraction to multiply with the size of the colorbar
extend	neither, both, min, max	Sets edges of the colorbar to be pointy if out-of-range values are present

6.3.2. Mesh Plots

Like contour plots, mesh plots are also two-dimensional plots that display three dimensions of information using *x, y,* coordinates and *z* for a color scale. However, mesh plots do not perform any smoothing and display data as-is on a regular grid. Since many satellite datasets are swath-based, irregularly spaced data needs to be re-gridded in order to display it as a mesh grid. In the code block below, let's compare how the MOPITT data looks using `pcolormesh` command with the previous example using `contour`. I made no other changes to the plot other than the call to the plot type.

```
fig, ax = plt.subplots()

co_plot = ax.pcolormesh(X_co, Y_co, co.transpose())
fig.colorbar(co_plot, orientation='horizontal')

ax.grid(True)

plt.show()
```

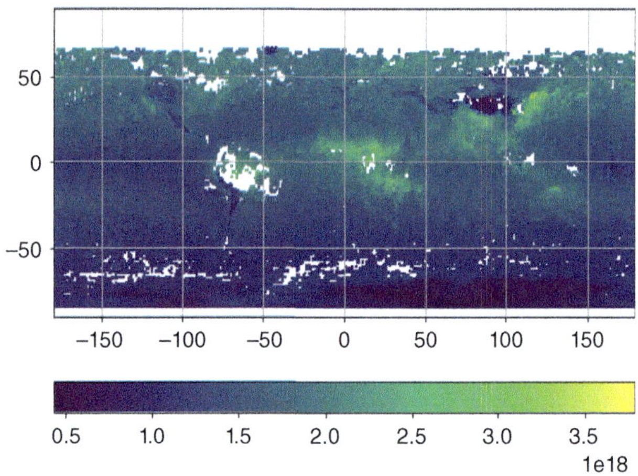

You might notice that there is more structure in the mesh plot than the filled contour. This is useful if you wish to examine fine structure and patterns. Remote sensing data are often noisy because atmospheric and surface conditions, such as clouds or snow, which can introduce errors into the retrieval. I usually prefer `pcolormesh` to `contourf`, but the latter is an appropriate choice depending on the application. Next, let's see if the `fires_count` data are easier to interpret in mesh form.

To convert the data to logscale in pcolormesh, I must use the norm= option to rescale the colors. I also added the optional vmin and vmax keywords to fix the color scale of the plot.

```
fig, ax = plt.subplots()

im = ax.pcolormesh(X_fires, Y_fires, fire_count_3D,
norm=colors.LogNorm(), vmin=1, vmax=1000)
fig.colorbar(im, orientation='horizontal', shrink=0.5,
pad=0.25)

ax.grid(True)

plt.show()
```

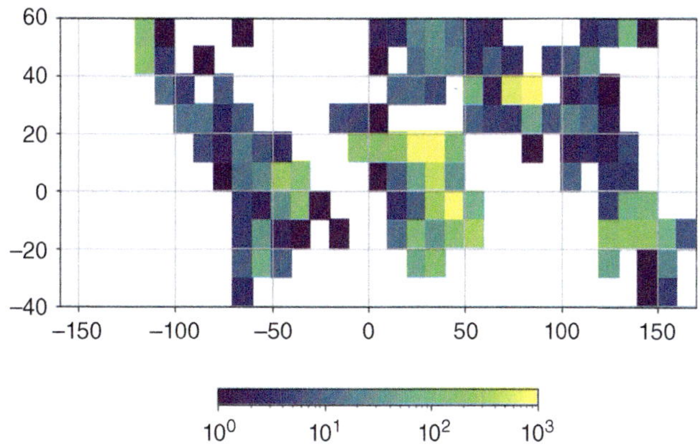

The above examples show a highly simplistic way of displaying three-dimensional data. The data in both fire_count plots show number of fires within a 10 × 10 degree latitude and longitude grid; however, the figures look quite different in contour and pcolormesh. The contour plot smooths over the features, which can more quickly communicate trends and patterns. In contrast, pcolormesh print a grid of the data and displays more detail. In the above example, the global distribution of fires varies significantly spatially, so smoothing the data is not advised. However, as you saw for monthly mean CO measurements, contour plots are an effective plot choice.

In both contour and mesh plot examples, a crucial element for interpreting the above plots is missing: I did not project the data on top of a map. In the next chapter, you will learn how to add in coastlines and other surface features to geolocated data.

6.4. Summary

In this chapter, you learned how to use matplotlib to create simple line and scatter plots. You made plots of one-, two, and three-dimensional data and covered some useful aesthetics, such as plot arrangement, layering data on the same figure, adding labels, and modifying color schemes. In the next chapter, you will work with satellite imagery that is geolocated on the map. This single addition to the plot greatly helps with data interpretation, making it perhaps the most important aesthetic decision for the figures.

References

ColorBrewer: Color Advice for Maps. (n.d.). Retrieved November 25, 2020, from https://colorbrewer2.org/

GOS_Components. (n.d.). Retrieved November 25, 2020, from https://www.wmo.int/pages/prog/www/OSY/Gos-components.html#upper

ncl_default color table. (n.d.). Retrieved November 25, 2020, from https://www.ncl.ucar.edu/Document/Graphics/ColorTables/ncl_default.shtml

OPeNDAP™. (n.d.). Retrieved April 5, 2020, from https://www.opendap.org/

Python strftime reference. (n.d.). Retrieved November 25, 2020, from https://strftime.org/

Rheingans, P. L. (2000). Task-based color scale design. In 28th AIPR workshop: 3D visualization for data exploration and decision making (Vol. 3905, pp. 35–43). International Society for Optics and Photonics. https://doi.org/10.1117/12.384882

Same stats, different graphs. (n.d.). Retrieved November 25, 2020, from https://www.autodesk.com/research/publications/same-stats-different-graphs Kovesi, P. (2015). Good Colour Maps: How to Design Them. ArXiv:1509.03700 [Cs]. Retrieved from http://arxiv.org/abs/1509.03700

Schultz, D. M. (2009). *Eloquent science: A practical guide to becoming a better writer, speaker, and atmospheric scientist*. Boston, Mass: American Meteorological Society.

Smith, N., R.B. Esmaili, C. D. Barnet, G. J. Frost, S. A. McKeen, M. K. Trainer, and C. Francoeur (2020). Monitoring atmospheric composition and long-range smoke transport with NUCAPS satellite soundings in field campaigns and operations. Presented at the 100th American Meteorological Society Annual Meeting, AMS. Retrieved from https://ams.confex.com/ams/2020Annual/webprogram/Paper368222.html

Wilkinson, L. (2005). *The grammar of graphics* (2nd ed.). New York: Springer-Verlag. Retrieved from https://www.springer.com/in/book/9780387245447

7

CREATING EFFECTIVE AND FUNCTIONAL MAPS

Earth observations are often communicated by projecting onto maps. This chapter provides a brief overview of projection types, how to convert between projections, and some common filtering options. In Python, one of the most common packages used to project datasets onto maps is Cartopy. Building on the previous chapter's tutorials on Matplotlib plotting, examples in this chapter show how Cartopy can be used to overlay maps onto plots, convert data between projections, and customize the scale of and content of map features. In addition, the xarray package streamlines some common plotting and map projections using both Cartopy and Matplotlib.

The package Cartopy add mapping functionality to Matplotlib. Cartopy provides an interface to obtain continent, country, and feature details to overlay onto your plot. Furthermore, Cartopy also enables you to convert your data from one map projection to another, which requires a cartesian coordinate system to the map coordinates. Matplotlib natively supports the six mathematical and map projections (Aitoff, Hammer, Lambert, Mollweide, polar, and rectilinear) and combined with Cartopy, data can be transformed to a total of 33 possible projections.

The examples in this chapter use the following packages that have been covered earlier: **NumPy, pandas, matplotlib.pyplot, netCDF4,** and **xarray.** I will introduce one new package, **Cartopy**, which is not included in the base installation of Anaconda. If you are working through these examples on your local computer, I recommend ensuring that these packages are installed (Appendix A.3).

Earth Observation Using Python: A Practical Programming Guide, Special Publications 75, First Edition. Rebekah B. Esmaili.

DOI: 10.1002/9781119606925.ch7

Finally, you will utilize the following datasets in the examples:

- **VIIRSNDE_global2018312.v1.0.txt.** The VIIRS Active Fire Product, which can help us see how many fires were detected from space on 2018 7 Nov. This dataset contains a set of latitude and longitude pairs.
- **JRR-AOD_v1r1_npp_s201808091955538_e201808091957180_c201808092049460.nc.** A netCDF file that contains Aerosol Optical Depth (AOD) retrieved from a Suomi NPP overpass on 2020 9 Aug. This dataset contains a swath of irregularly spaced two-dimensional data.
- **RDEFT4_20200131.nc.** A sea ice thickness dataset from ESA's CryoSat-2, which can be usefully visualized on maps of the North and South pole.
- **OR_ABI-L1b-RadM1-M3C13_G16_s20182822019282_e20182822019350_c20182822019384.nc.** A GOES-16 ABI mesoscale image of Hurricane Michael from 2018 09 Oct. The data are in the geostationary satellite projection.
- **GLM-L2-LCFA_2018_282_20_OR_GLM-L2-LCFA_G16_-s20182822000200_e20182822000400_c20182822000427.nc.** A global lightning dataset
- **sst.mnmean.nc.** A netCDF file that contains the mean Sea Surface Temperature (SST) for the month of Nov 2018.

7.1. Cartographic Projections

7.1.1. Geographic Coordinate Systems

Many remote sensing datasets include geolocation in the form of **latitude** and **longitude** to define the position of an observation. Latitude is represented as a series of parallel lines that extend from pole to pole (90°N to 90°S). The distance that a degree of latitude represents is the nearly same across the entire Earth. Longitude intersects latitude at right angles. At the equator, each degree of longitude is nearly spaced the same as each degree of latitude (which is 111 km and 110 km, respectively), but degrees longitude shorten significantly moving poleward, starting at 111 km at the equator, shrinking to 78 km at 45°, and becoming infinitesimally small at 90°N and 90°S. Longitude is typically expressed from –180 (180°W) to 180 (180°E) or, alternatively, in coordinates from 0° to 360°. Latitude and longitude simplify the Earth's surface to be a perfect oblate spheroid, ignoring the elevation or altitude of the point.

Another geographic coordinate system is the Universal Transverse Mercator (UTM), which expresses data as a numbered zone (spanning 0–60) on a Mercator projection, and within them a northing and easting coordinate in meters from the bottom left corner of the zone. An advantage of this system is that the distance between points is very easy to calculate. Landsat and some high-resolution datasets like AVIRIS use UTM. While I want to make you aware that other systems exist, examples in this chapter focus on the latitude–longitude coordinate system.

This book uses the World Geodetic System 1984 (WGS84) coordinate reference system. WGS84 defines the origin as Earth's center of mass, the zero longitude line as the Greenwich meridian, the zero latitude line as the equator, and several standard parameters to describe the Earth's geometry. These definitions, in turn, determine the position of an observation in its geographic coordinate system. WGS84 is the default option for many projections in Cartopy, but the projections can often accommodate other systems and definitions.

7.1.2. Choosing a Projection

To create a plot, spherical coordinates latitude and longitude must first be **projected**, or mathematically transformed, onto a two-dimensional, gridded plane. This is because plots and figures are usually viewed on a flat screen or printed on paper, although three-dimensional and other projection surfaces are increasingly being used as technology advances. While an infinite number of projections are mathematically possible, out of practicality the map projections are based on three regular shapes: cylindrical, conic, or plane/azimuthal, which are shown in Figure 7.1.

These maps, along with some features, are described in Table 7.1. Every map projection will inherit some form of distortion. For instance, some maps appear flat, as if they have been unrolled onto a table (such as Mercator). These flattened projections preserve the scale of the latitude and longitude coordinates, but not the distance between them (and therefore, area). In midlatitudes, the distance between a line of latitude and longitude is roughly equal. However, as you move poleward,

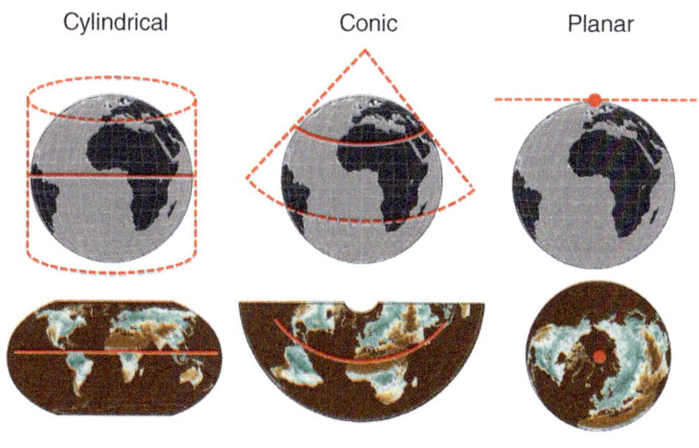

Figure 7.1 Most varieties of map projections are cylindrical, conic, or planar.

Table 7.1 Map Projection Shapes, Examples, and Regions of Minimal Distortion

Name	Description	Example projections	Minimal distortion
Cylindrical	Often "flat" in shape but also can take on oval shapes with parallel longitude circles.	Mercator, Plate Carrée/equal angle	Tropical regions
Conic	Often fan and lantern-shaped plots but can take some unusual shapes.	Lambert conformal conic	Midlatitudes
Planar, azimuthal	Picks a single point, as opposed to a latitude or longitude circle as the area of focus.	Orthographic, Lambert azimuthal equal-area	Polar regions

the spacing between longitude lines is significantly smaller than latitude lines. This will inflate the area of northern land masses and features relative to those closer to the equator.

In Figure 7.2, darker regions show which areas have the most distortion while lighter colors have least distortion. You can modify some of the map parameters to minimize distortion over a region of interest. For example, the default origin in Cartopy on planar-type maps may be 0°,0°. However, by changing the map origin coordinates (for instance, to 23.6° S, 46.6° W if you are interested in data over Sao Paolo, Brazil), you can shift the regions with the most distortion by re-centering your plots.

When designing your maps, you must choose what quantitative feature you most want to preserve: area, shape, distance, or direction. Below is a short definition of each:

- **Area.** We preserve area across all regions to fix the scale regardless of location. A quarter near the equator is the same spatial area on a map as a quarter in the midlatitudes and poles.

Figure 7.2 Each projection type has different regions of distortion. Lighter colors indicate less distortion, while darker colors show more.

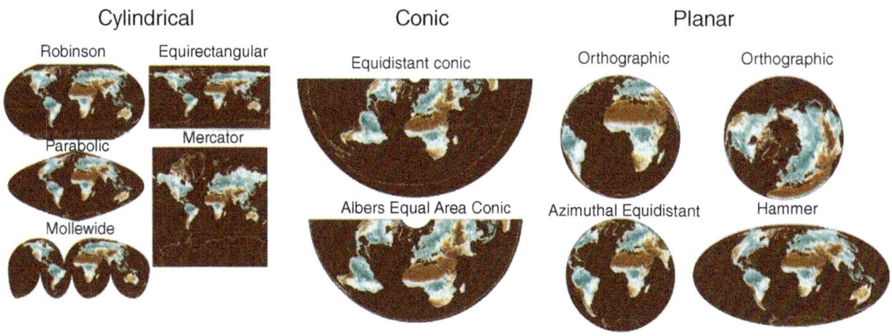

Figure 7.3 Examples of the types of maps within each projection category.

- **Shape.** Shape has to do with the skewedness of the original object. For instance, if a continent is preserved after the projection, the essential shape of the continent might look different.
- **Scale or distance.** Preserves the "true scale" with respect to some central area.

Generally, only one of the requirements can be met at the expense of the other. So in addition to their fundamental shape, map projections can be subclassified according to what cartographic features they preserve (Figure 7.3).

- **Equidistant.** Plate Carrée, equirectangular; preserves distance from a point or line; typically associated with airline flight maps.
- **Equal-area.** Albers conic, Lambert azimuthal equal-area; preserves area, which then distorts shapes.
- **Conformal.** Orthographic (plane style), Lambert conformal conic, Mercator (not equal angle or equidistant); preserves angles locally. Best for large-scale maps.

7.1.3. Some Common Projections

The above discussion is not meant to overwhelm you with choices, but rather, give you some background to better understand the philosophy of the map projections. The remainder of this section explores some of the common projections in more detail. For more detail, the references at the end of the chapter provide a more through discussion of these projections and map projections in general.

7.1.3.1. Plate Carrée

Plate Carrée (French for "flat square") is a type of equirectangular projection that has minimal distortion along the equator, which falls into the equidistant map category. It is a common projection because it is easy to plot a pixel on these maps – the longitude and latitude coordinates are equal to the x and y geometric coordinates, respectively. Thus, no further transformation needs to take place. This was especially advantageous before the availability of personal computers, when more complex maps would take longer to create.

7.1.3.2. Equidistant Conic

This projection is conic, and, like Plate Carrée, is equidistant. The scale is preserved along all latitude circles. The central point will determine which longitude circles have preserved scale.

7.1.3.3. Orthographic

Orthographic plots are conformal, usually centered at 0°N, 0°E, but can also be centered at the poles. This plot is advantageous because it looks spherical, and data look how we would intuitively expect them to if we were in space. This projection is useful for geostationary data, which continuously monitors one region of the Earth. Using many other map projections, polar data are poorly represented but can be the least distorted using orthographic. Significant angular distortion occurs at the edges, where pixels will appear larger than at the center.

7.2. Cylindrical Maps

7.2.1. Global Plots

For the following set of examples, I will examine the VIIRS Active Fire Product, which is produced by NOAA for daily fire and smoke analysis. The data contain fire positions, confidence value, fire mask, brightness temperature, fire radiative power, and other metadata and in comma-delimited tables. Continuing the examination of the California Camp Fire from Chapter 6, I will look at the active fires for 8 November 2018, when the main fire started.

Below I import NumPy, Pandas, and matplotlib, and since I will be adding map overlays and projects, I also import `cartopy.crs`, which stands for Coordinate Reference Systems, and give it the alias `ccrs`. As an option, I am increasing the figure and font size so that they are easier to read.

```
import pandas as pd
import numpy as np
from cartopy import crs as ccrs
import matplotlib.pyplot as plt

plt.rcParams.update({'font.size' : 16, 'figure.figsize' :
[12, 6]})
```

Next, I import the VIIRS dataset and use the `head` command to inspect the few lines:

```
fires = pd.read_csv("data/VIIRSNDE_global2018312.v1.0.txt")
fires.head()
```

	Lon	Lat	Mask	Conf	brt_t13(K)	frp(MW)	line	sample	Sat	YearDay	Hour
0	27.110006	30.769241	8	52	302.877533	5.814295	242	1735	NDE	2018312	1
1	26.083252	30.534357	9	100	332.959717	24.340988	301	1620	NDE	2018312	1
2	34.865997	28.162659	8	38	301.165985	6.107953	396	2589	NDE	2018312	1
3	34.872623	28.161121	8	71	307.277985	9.287819	396	2590	NDE	2018312	1
4	34.865070	28.158880	8	39	301.227783	6.001442	402	2590	NDE	2018312	1

Each row has a latitude and longitude coordinates pair; altogether, this DataFrame is an array of points. These two coordinates are plotted on a global map using `plt.scatter`. Cartopy integrates with the matplotlib plots, so the code below will build on what you learned in Chapter 6. Below, there are three new lines of code in the second block. The first line defined the projection of the axes to be Plate Carrée (which is mapped to a variable named `ax`). In the second new line, the coastline feature is added using `ax.coastline`. Setting the axes to global (using `ax.set_global`) ensures that I can see the coastlines of the entire Earth. The remainder of the code plots a scatter plot of the latitude and longitude coordinates of the fire.

```
fig = plt.figure(figsize=[20,20])

ax = plt.subplot(projection=ccrs.PlateCarree())
ax.coastlines()
ax.set_global()

plt.scatter(fires['Lon'], fires['Lat'])
plt.show()
```

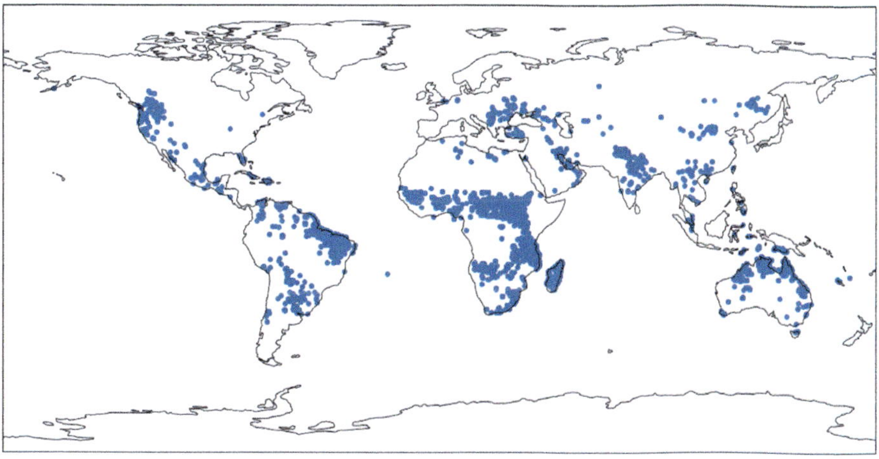

7.2.2. Changing Projections

Data in one projection can be transformed into another projection. Fortunately, Cartopy will perform all the mathematical calculations for supported projections. In the next example, I will switch from Plate Carrée to Lambert Azimuthal Equal Area. I must define the projection twice. In the `plt.subplots` line, I must define the **to coordinates** (`ccrs.LambertAzimuthalEqualArea`), which is how I want to axes to show the data. In the `ax.scatter` line, I use the transform keyword argument in scatter to define the **from coordinates** (Plate Carrée), which are the coordinates that the data formatted for.

For aesthetics, I added the gridlines to the plot using the `ax.gridlines` command. Let's give it a try below:

```
fig = plt.figure(figsize=[20,20])

ax=plt.subplot(projection=ccrs.LambertAzimuthalEqualArea())

ax.coastlines()
ax.gridlines()
ax.set_global()

plt.scatter(fires['Lon'], fires['Lat'],
transform=ccrs.PlateCarree())
plt.show()
```

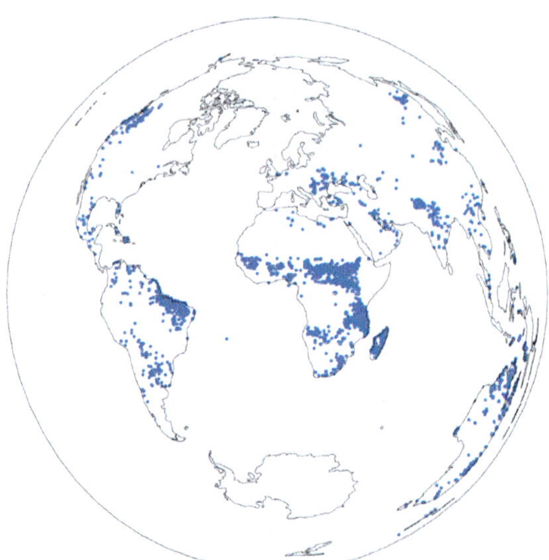

7.2.3. Regional Plots

Since I am specifically interested in a Californian wildfire, I can change the axes x and y ranges to zoom in between 125°W–120°W and 38°N to 44°N. I define these longitude and latitude ranges below in the extent variable:

```
extent = [-125, -120, 38, 44]
```

Since I am zooming in, I also want to increase the higher resolution coastline resolution. Cartopy has coastline resolutions of 110 m (default), 50 m, and 10 m. Note that the 50 m and 10 m coastlines need to be downloaded from the internet the first time they are used. The feature resolution of the coastline is increased by including the `50 m` option. If you are not connected to the internet, you can remove the `50 m` specification and the coastlines will be in the default 110 m like in the previous example.

Map projection names can be very long, such as `LambertAzimuthal-EqualArea`, which is 25 characters. I personally think the code is easier to read if I assign the projection names to variables, which I do in the code block below. In the next example, the projection I am converting to is the same as the projection I am projecting from, so this is somewhat unnecessary. I still organize my code this way because if I want to change the plot projections, it is trivial to update one variable. Otherwise, I would have to search my code and find all the references that need the projection name changed.

```
to_proj = ccrs.PlateCarree()
from_proj = ccrs.PlateCarree()
```

In the previous example, you will notice that there are no latitude and longitude labels in the plots, but they are often helpful when examining plots and sometimes necessary for publication. Cartopy has a `gridliner` feature to nicely format latitude and longitude in Plate Carrée or Mercator projections only. At the time of writing, the developers are working on improving supporting more map projections.

```
from cartopy.mpl.gridliner import LONGITUDE_FORMATTER,
    LATITUDE_FORMATTER
from matplotlib import ticker
```

To add the labels, import `ticker` from matplotlib, which can both locate the ticks and apply formatting. In Section 6.4.1, I used `ticker` to format the fire data to log space in a contour plot. `Ticker` can also be used to locate the coordinates. Specifically, I define the coordinates that I wish to locate using `np.arange` below. For example, I want to space my longitude coordinates every 1.5° and my latitude coordinates every 1°. You can customize the spacing as you wish.

```
lonLabels = np.arange(-180, 180, 1.5)
latLabels = np.arange(-90, 90, 1)
```

Now I can assemble the plotting code. The first block of code looks like the past examples: I initialize the figure, add the axes and projections, and overlay the data. One difference is that I change the color and size of the scatter points to scale with the fire radiative power. As a result, the plot will now need a color bar (`plt.colorbar`).

Adding the gridlines and labels is a little more complex, so each comment block is numbered and does the following:

1. Updates `ax.gridlines` with `draw_labels=True`. The default option in Cartopy is `draw_labels=False`, which is why the previous plots did not have grid lines or labels across them. This is assigned to the variable `gl`, which I use to apply further gridlines and label customizations.
2. Calls the matplotlib fixed locator and updates `xlocator` and `ylocator` attributes with the default tick locations with those I defined in `lonLabels` and `latLabels`.
3. Updates Cartopy's the `xformatter` and `yformatter` to overwrite the default labels (e.g., -130, -125, -120) with "nice" labels (e.g., x coordinates -130, -125, -120 become 130W, 125W, 120W, etc.) using Cartopy's `LONGITUDE_FORMATTER` and `LATITUDE_FORMATTER`.
4. Sets the top and right labels to `False`. This is optional and a matter of personal preference. If you choose you can delete these lines and all the axes will have latitude and longitude labels.

```
fig = plt.figure()
ax = plt.subplot(projection=to_proj)
ax.coastlines('50m')
ax.set_extent(extent)

plt.scatter(fires['Lon'], fires['Lat'], c=fires['frp(MW)'],
    s=fires['frp(MW)'], transform=ccrs.PlateCarree())

cbar = plt.colorbar()
cbar.ax.set_title("Fire Radiative Power (MW)",
    rotation='vertical', x=-0.5)

# 1. Maps the gridlines to the variable gl
gl = ax.gridlines(crs=to_proj, draw_labels=True)
```

```
# 2. Adds two attributes to gl, which are xlocator and
ylocator
gl.xlocator = mticker.FixedLocator(lonLabels)
gl.ylocator = mticker.FixedLocator(latLabels)

# 3. Changes labels to show degrees North/South and
East/West
gl.xformatter = LONGITUDE_FORMATTER
gl.yformatter = LATITUDE_FORMATTER

# 4. Removed labels from top and right side
# comment out if you want to include
gl.xlabels_top = False
gl.ylabels_right = False

plt.show()
```

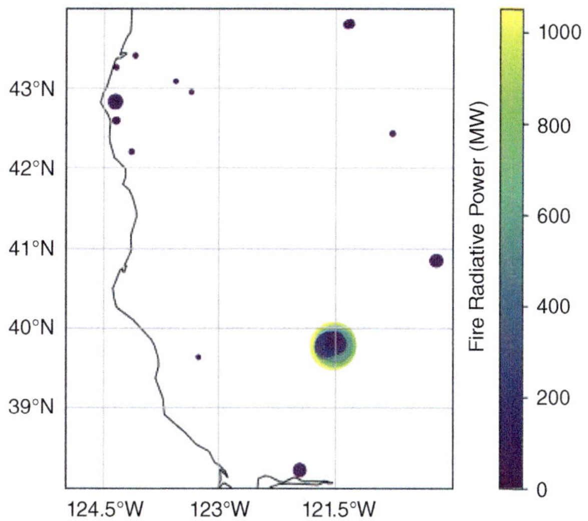

In general, I strongly recommend including the coordinate labels in your plots because they are easier to interpret. However, to keep code blocks short in this text, I will omit them for most examples.

7.2.4. Swath Data

The above example mapped an array of coordinates using the scatter plot function within matplotlib. Satellite datasets are often stored as two-dimensional arrays, or mesh grids. The data are functions of longitude and latitude, which means three values need to be displayed. Often, satellite data are very high resolution, so data are often stored in `swaths`, or small strips of data, such as all the observations collected over a 30-second period. The swaths can then be pieced together to make a complete picture of the globe.

In the next example, I will look at the Aerosol Optical Depth (AOD) over the California during the Camp Fire and make a contour plot. In Section 6.4, I described the differences between contour plots and meshplots. Contour plots are visualizations that show the concentration of values and help smooth data into levels, either by default within matplotlib or using user specific ranges. Contour plots are useful for data with high spatial correlation, such as air temperature or AOD. However, keep in mind that if a dataset has large variability between pixels, the contour levels can smooth the data too much and you will not be able to see small scale features. In some cases, this can dramatically alter the interpretation of the results.

The process for making contour plots for remote sensing data builds on all the skills you have learned so far. First, I need to import the data, which is commonly in netCDF4 or HDF format (Chapter 5):

```
from netCDF4 import Dataset

fname = 'data/JRR-AOD_v1r1_npp_s201808091955538_ e2018080
91957180_c201808092049460.nc'
file_id = Dataset(fname)
```

If you inspect the timestamps in the filename (the numbers following `_s` and `_e`), this file contains a single swath of data that was collected over two minutes. Below, I will pull variables related to the AOD and to show the movement and range of the smoke plume from the fires. This dataset is derived from VIIRS, which is a sensor on the Suomi NPP polar orbiting satellite (Section 1.1.1).

As a reminder, you can inspect the contents of the file by printing the available variables inside the file:

```
file_id.variables.keys()
odict_keys(['Latitude', 'Longitude', 'StartRow',
'StartColumn', 'AOD550', 'AOD_channel', 'AngsExp1',
'AngsExp2', 'QCPath', 'AerMdl', 'FineMdlIdx',
'CoarseMdlIdx', 'FineModWgt', 'SfcRefl', 'SpaStddev',
'Residual', 'AOD550LndMdl', 'ResLndMdl', 'MeanAOD',
'HighQualityPct', 'RetrievalPct', 'QCRet', 'QCExtn',
'QCTest', 'QCInput', 'QCAll'])
```

Of these variables, I need `AOD550` along with `Latitude` and `Longitude` so that I know where to plot the data. Before I can make a plot, I need to know the dimensions of the data. Below I print the shape of the data, and learn that the data are two-dimensional and have the same dimensions:

```
print(file_id.variables['AOD550'].shape,
      file_id.variables['Latitude'].shape,
      file_id.variables['Longitude'].shape)
(768, 3200) (768, 3200) (768, 3200)
```

I import the AOD, latitude, and longitude data into variables into a two-dimensional NumPy array by using square brackets `[:,:]`.

```
aod = file_id.variables['AOD550'][:,:]
lat = file_id.variables['Latitude'][:,:]
lon = file_id.variables['Longitude'][:,:]
```

Rather use the default matplotlib contour levels, I want to manually scale the data. Using the `valid_range` attribute, I can see that AOD ranges from –0.05 to 5:

```
print(file_id.variables['AOD550'].valid_range)
[-0.05  5.  ]
```

However, in nature, most AOD values fall under 2.0 except in heavily polluted environments. Below I define the range from 0 to 1.8 with increments of 0.1 (this is where the contour lines will be drawn). Note that in Python, the upper value of the range is excluded, and the variable `levs` stops at 1.7. If I wanted to include 1.8, I need to increase the end value by one increment (1.9).

```
levs = np.arange(0, 1.8, 0.1)
levs
array([ 0. ,  0.1,  0.2,  0.3,  0.4,  0.5,  0.6,  0.7,  0.8,  
0.9,  1. ,
        1.1,  1.2,  1.3,  1.4,  1.5,  1.6,  1.7])
```

Finally, I plot the data using my projection of choice (Plate Carrée) over California (using `extent`). Recall from Section 6.3.1 that `plt.contourf` has an parameter input order of *x, y, z*. You can alternatively explicitly state which variable such as using `x=, y=, z=`. All three variables must be two-dimensional and have the same shape. The fourth keyword contains the levels. The `extend='both'` allows values outside of my range (which is 0 to 1.7, based on my definition of levs) to be included in the plot, but will have the maximum or minimum color value. If `extend='neither'`, an AOD of 2.0 will be missing, for example. The colorbar was have arrows at the endpoints to show that the color scales include out of range data. In addition to displaying the contours, I include ax.scatter in this block of code so that I can overlay the active fire regions onto AOD.

```
fig = plt.figure(figsize=[12, 12])
ax = plt.subplot(projection=ccrs.PlateCarree())

ax.coastlines('50m')

extent = [-125, -120, 38, 44]
ax.set_extent(extent)

x1 = ax.contourf(lon, lat, aod, levs, extend='both')

fig.colorbar(x1, extend='both', orientation="horizontal",
    fraction=0.05)

# Adds the active fire scatter plot on top
ax.scatter(fires['Lon'], fires['Lat'], color='red', s=50)

plt.show()
```

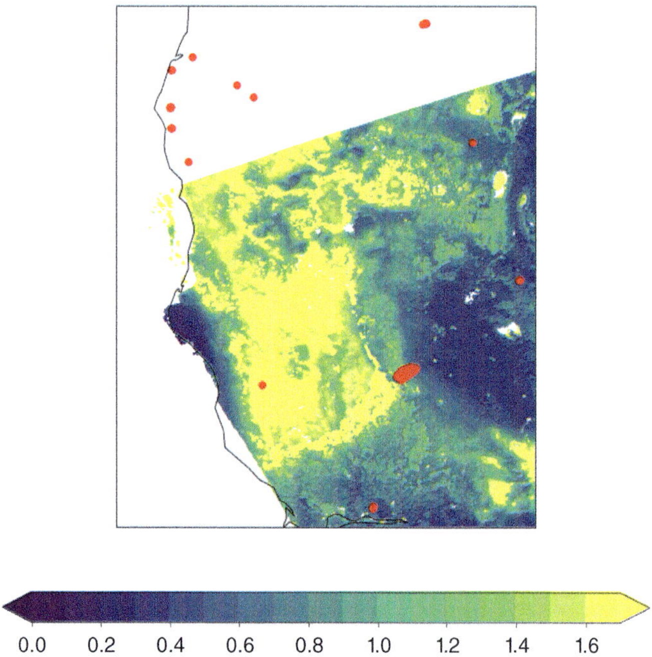

7.2.5. Quality Flag Filtering

Remote sensing datasets commonly have **quality flags** to inform the user of uncertainly in the results. Filtering low-quality data from your maps can help remove unphysical artifacts and may make them more effective visuals. I discussed some of the details back in Section 3.2.3. If your model or analysis is highly sensitive to the inputs, you should only use the **best-quality data**. The meaning and value of quality flags can vary between datasets, so you may need to consult user guides for each dataset to clearly understand the meaning of the quality flag. AOD quality flags for VIIRS is specified on an integer scale with the following meaning:

```
0: Best
1: Medium
2: Low
3: No retrieval
```

To remove all values that are not 0 (best), I can use a mask (Section 4.5) to filter out the remaining values. First, I import the quality flag variable into a NumPy array.

```
# Import quality flag
quality_flag = file_id.variables['QCAll'][:,:]

# Keep all but the "best" quality using masked arrays
maskHQ = (quality_flag != 0)
aodHQ = np.ma.masked_where(maskHQ, aod)
```

Suppose that I am curious about how much data are removed when I apply the quality flag. I can use NumPy's `.count()` to compute the number of `True` cells in the mask:

```
(aod.count()-aodHQ.count())/aod.count()
0.33269544384282035
```

So, roughly 33% of cells removed from this swath. Note that the percentage will vary from swath to swath.

To compare the unmasked to the masked data, place the two plots together side by side (Section 6.3.6). Aside from adding titles to the two plots, the remainder of the code is unchanged from the previous example.

```
# Top plot
fig = plt.figure()
```

```
upper_axis = plt.subplot(2,1,1,
projection=ccrs.PlateCarree())
upper_axis.set_title("All Quality")

upper_axis.coastlines('50m')

upper_fig = upper_axis.contourf(lon, lat, aod, levs,
extend='both')
fig.colorbar(upper_fig, ax=upper_axis, extend='both')

# Bottom plot
lower_axis = plt.subplot(2,1,2,
projection=ccrs.PlateCarree())
lower_axis.set_title("High Quality")

lower_axis.coastlines('50m')

lower_fig = lower_axis.contourf(lon, lat, aodHQ, levs,
    extend='both')
fig.colorbar(lower_fig, ax=lower_axis, extend='both')

plt.show()
```

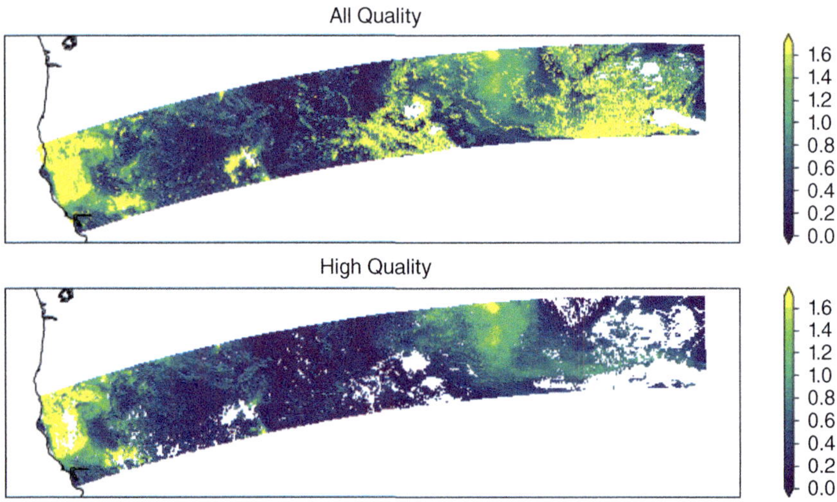

Many of the very high values have been removed after I applied quality control. For this swath, this is likely due to the presence of clouds. AOD makes retrievals using visible channels, which like the human eye, cannot "see" through clouds. When smoke plumes are very thick, this can also interfere with the retrieval. In this case, smoke from the wildfires is likely inhibiting accurate retrievals closer to the coastline. By filtering the data, it is easier to visually distinguish regions that have high AOD values rather than those that are detecting clouds.

7.3. Polar Stereographic Maps

Sea ice is more common in the polar regions, so planar maps that are centered on the North or South Poles are useful for displaying this data. In the following example, I will display ESA's CryoSat-2 sea ice thickness dataset on a stereographic map. Below, I import the data using **xarray**. If you prefer to import the data using the **netcdf4** package, refer to the previous example. First, I open the file using the `open_dataset` command in xarray, save it to a variable called ice, and print the header information:

```
import numpy.ma as ma
import xarray as xr

fname = "data/RDEFT4_20200131.nc"
ice = xr.open_dataset(fname)

print(ice)
```

```
<xarray.Dataset>
Dimensions:                (x: 304, y: 448)
Dimensions without coordinates: x, y
Data variables:
    sea_ice_thickness  (y, x) float32 ...
    snow_depth         (y, x) float32 ...
    snow_density       (y, x) float32 ...
    lat                (y, x) float32 ...
    lon                (y, x) float32 ...
    freeboard          (y, x) float32 ...
    roughness          (y, x) float32 ...
    ice_con            (y, x) float32 ...
Attributes:
    Title:           CryoSat-2 sea ice thickness and
ancillary data
    Abstract:      ... monthly averaged Arctic sea ice
thickness ...
```

```
    Projection:    CryoSat-2 elevation data have ...
    Institution:    NASA Goddard Space Flight Center
    PI_name:    For questions related to this data set ...
    References:    A description of the primary
methodology ...
```

From the output above, I need the `sea_ice_thickness`, `lat`, and `lon` variables to make the plot. When I check the metadata, I can see there is no `_FillValue` attribute:

```
print(ice['sea_ice_thickness'])
<xarray.DataArray 'sea_ice_thickness' (y: 448, x: 304)>
[136192 values with dtype=float32]
Dimensions without coordinates: y, x
Attributes:
    units:            Meters
    long_name:    Sea ice thickness
```

However, the mean value is negative (Section 5.2). Given that the metadata says that the units are in meters, it should be suspicious that the data have negative values.

```
ice['sea_ice_thickness'].mean()
<xarray.DataArray 'sea_ice_thickness' ()>
array(-9044.761, dtype=float32)
```

Printing the minimum value shows that the fill value is probably -9999.0.

```
print(ice['sea_ice_thickness'].min())
<xarray.DataArray 'sea_ice_thickness' ()>
array(-9999., dtype=float32)
```

Although it would be best if all datasets had a defined fill value attribute, I can usually do some detective work in cases where the fill value is absent. Since the fill value was not defined in the file, I now also must manually mask these values out. I can use the `where` command, which is originally from NumPy but can also be called in DataArrays. The command below applies the mask to all the variables in the dataset, including the `sea_ice_thickness`:

```
ice_masked = ice.where(ice['sea_ice_thickness'] != -9999.0)
```

Now the data are ready to make the plot. At the top of the code, I have two variables that define what projection the data was originally in (`from_proj`) and the new projection that I want to map it to (`to_proj`).

```
to_proj = ccrs.NorthPolarStereo()
from_proj = ccrs.PlateCarree()
```

Below, I use pcolormesh (Section 6.3.2) to display the sea ice data. Also, adding gridlines and the set_extent options to the plot make it easier to see.

```
plt.figure(figsize=[10,10])
ax = plt.subplot(projection=to_proj)
ax.coastlines('50m')
ax.gridlines()

ice_plot = ax.pcolormesh(ice['lon'], ice['lat'],
    ice_masked['sea_ice_thickness'], transform=from_proj)
plt.colorbar(ice_plot)

ax.set_extent([-180, 180, 60, 90], crs=from_proj)
plt.show()
```

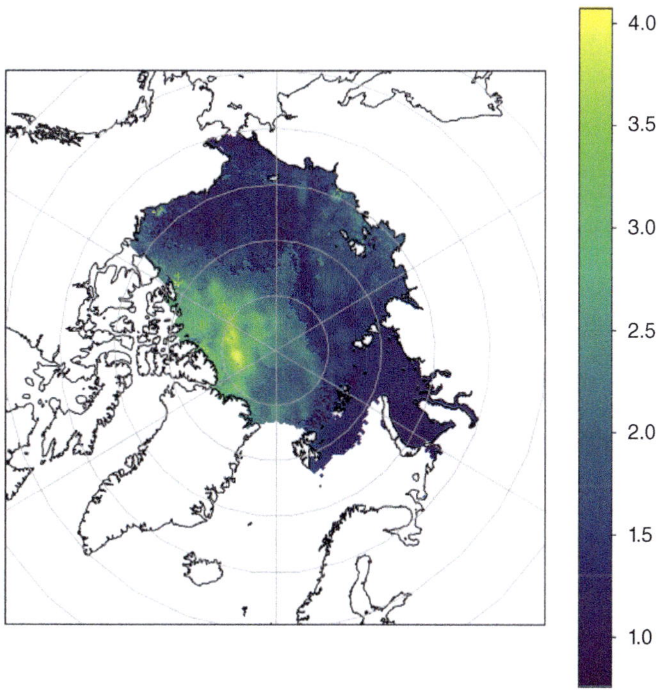

As you can see, the steps for working with different projections are similar, especially if the coordinates are included in the file. In the next section, I will work with geostationary data, which does not contain the coordinates.

7.4. Geostationary Maps

Geostationary orbits match the rotation of the Earth, so they can provide continuous imaging over the same domain. The geostationary projection is a special case of the planar map, where the central point is on the equator. In the next section, I will use imshow to display a 10.3 μm image of Hurricane Michael on 9 October 2018 from the GOES-16 ABI instrument. Like pcolormesh, imshow is useful for showing the finer spatial details in a plot. The main difference between imshow and pcolormesh is the pixel size assumption. In imshow, the pixels are assumed to be equal sized and regular. pcolormesh assumes the pixels are rectangular and they can have a nonregular (nonuniform) grid. imshow can be used with that data to provide a consistent spacing. In the example below, the pixels have viewing zenith angles less than 90° so the distortion is minimal.

First, I import a GOES-16 ABI (Section 1.1.1) mesoscale image. The naming scheme is similar to VIIRS, but the dates are in Julian day (defined as the number of days since 1 January of that year), not year, month, and day format. So 9 October 2018, is Julian day 282:

```
fname = 'data/goes-meso/michael/OR_ABI-L1b-RadM1-M3C13_G16_
s20182822019282_e20182822019350_c20182822019384.nc'
file_id = Dataset(fname)
list(file_id.variables)
['Rad', 'DQF', 't', 'y', 'x',
'longitude_of_projection_origin',
'perspective_point_height', ...]
```

I printed out some of the variables inside the file. If you inspect the whole list, you will find that there are no latitude and longitude coordinates; this is because geostationary data uses a special projection grid to conserve file space. For example, in GOES ABI L1b data, the coordinates are stored in variables called x and y. Although the letters *x* and *y* may look like rectangular cartesian coordinates, in this dataset they are the azimuth and elevation angles in a spherical-coordinate system. And if you print x and y, you can see that they do not look like latitude and longitude coordinates at all. Cartopy can help transform these spherical coordinates into geographic coordinate values.

```
file_id.variables['x'][0:10], file_id.variables['y']
[0:10]
(masked_array(data = [-0.04424  -0.044184 -0.044128
```

```
-0.044072 -0.044016 -0.04396  -0.043904
  -0.043848 -0.043792 -0.043736],
             mask = False,
       fill_value = 1e+20),
 masked_array(data = [ 0.08848      0.088424     0.08836801
0.088312     0.088256     0.0882
   0.088144     0.08808801   0.088032     0.087976   ],
             mask = False,
       fill_value = 1e+20))
```

Using the geostationary projection requires some extra work, but the steps can be applied to other geostationary imagers, such as AHI or SEVIRI from Himawari and Meteosat, respectively. Following are the available projection variables:

```
proj_var = file_id.variables['goes_imager_projection']
print(proj_var)
```

```
<class 'netCDF4._netCDF4.Variable'>
int32 goes_imager_projection()
    long_name: GOES-R ABI fixed grid projection
    grid_mapping_name: geostationary
    perspective_point_height: 35786023.0
    semi_major_axis: 6378137.0
    semi_minor_axis: 6356752.31414
    inverse_flattening: 298.2572221
    latitude_of_projection_origin: 0.0
    longitude_of_projection_origin: -75.0
    sweep_angle_axis: x
unlimited dimensions:
current shape = ()
filling on, default _FillValue of -2147483647 used
```

There are four required parameters to create the geostationary projection (Figure 7.4). GOES-16 orbits the Earth at a height of roughly 35.8 km (`perspective_point_height`) from the surface of the Earth. From this vantage point, the GOES-16 ABI can see roughly 40% of the Earth surface. The center of the GOES-16 ABI's field of view is at 0°N, 75°W, which is also called the **subsatellite point** (`longitude_of_projection_origin`). The Earth more closely approximates an oblate spheroid in shape, not a sphere, because radius at the equator is roughly 21.4 km longer than the radius at the poles. In geometry, these parameters are respectively known as the **semi major axis** (`semi_major_axis`) and the **semi minor axis** (`semi_minor_axis`).

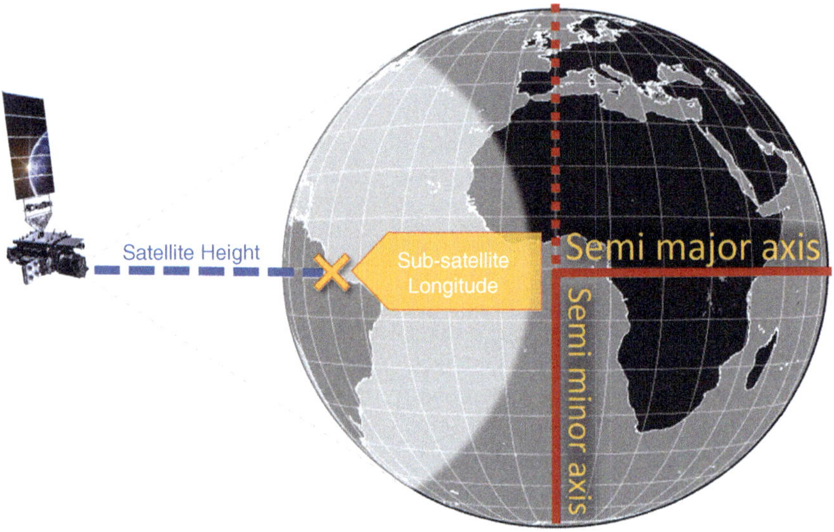

Figure 7.4 Imager projections from Geostationary Satellites.

To specify the GOES-16 ABI project in Cartopy, perform the following steps:

1. Import the `x`, `y`, and radiance (`Rad`) variables from the file.
2. Import the four projection parameters: `perspective_point_height`, `semi_major_axis`, `semi_minor_axis`, and `longitude_of_projection_origin`.
3. Multiply *x* and *y* times the satellite height to rescale the floats to the original pixel size.

   Steps 1–3 are performed in the code block below:

```
# Define the satellite height and central longitude for
plots
# Can vary depending on the geo satellite
sat_height = proj_var.perspective_point_height
```

4. Define the plane parameters that the data will be projected to. I will use `ccrs.Globe` to create a globe with the semi-major axis and semi-minor axis coordinates to approximate the Earth's geometry.

```
# Define the globe projection
semi_major = proj_var.semi_major_axis
semi_minor = proj_var.semi_minor_axis

globe = ccrs.Globe(semimajor_axis=semi_major,
    semiminor_axis=semi_minor)
```

5. Define satellite position using the central longitude. I do not need the central latitude because all geostationary satellites are centered at the equator. Thus, Cartopy sets the default value for `central_latitude=0`. I can then define the Geostationary grid using `central_longitude` and the `sat_height` with the globe I defined in the previous step using Earth's axis parameters.

```
central_lon = proj_var.longitude_of_projection_origin

crs = ccrs.Geostationary
(central_longitude=central_lon,
    satellite_height=sat_height, globe=globe)
```

Next, specify how the data are displayed using `imshow`. In the dataset I imported, the *x* and *y* values are scaled to a smaller set of decimal floating-point values to reduce the file size. I can define the extent by multiplying the *x* and *y* times the satellite height to convert the angle positions in radians to distance positions in meters from the Greenwich Meridian (CGMS, 2013).

```
# Multiply the x, y coordinated by satellite height to
# get the pixel position

X = file_id.variables['x'][:] * sat_height
Y = file_id.variables['y'][:] * sat_height
```

However, when I printed the *x* and *y* values earlier using `file_id.variables[...]`, you can see that *x* is increasing with each array index, while *y* is decreasing. In `imshow`, the default origin is the upper-left corner (`origin='upper'`). If I try to plot the data as-is, the image will be flipped since the *y* data are defined from the lower left. To fix this, I will explicitly tell `imshow` what the corners of the data are:

```
imgExtent = (X.min(), X.max(), Y.min(), Y.max())
```

Finally, I can display the plot by passing the defined geostationary projection (`projection=crs`), defining the coastlines (which I make orange to help distinguish from the underlying data), and the extent and origin parameters into `imshow`. As I mentioned earlier, imshow assumes the pixels are all the same size, so it does not require the *x* and *y* coordinates like `pcolormesh`.

```
fig = plt.figure(figsize=(10,10))

ax = plt.subplot(projection=crs)
ax.coastlines('10m', color='orange', linewidth=2)
```

```
ax.imshow(rad, origin='upper', cmap='gray_r',
extent=imgExtent)

plt.show()
```

Data that are not in the geostationary projection can also be overlapped on the same plot when transformed to the same projection. Below, I import GOES-16 geostationary lightning mapper (GLM) to see where lightning is in the hurricane, which indicates intensification.

```
glmfname = 'data/GLM-L2-LCFA_2018_282_20_OR_GLM-L2-LCFA_
G16_s20182822000200_e20182822000400_c20182822000427.nc'
file_id_glm = Dataset(glmfname)
file_id_glm.variables.keys()
[..., 'event_lat', 'event_lon', 'event_energy',
  'event_parent_group_id', 'flash_area'...]
```

GLM has multiple, one-dimensional coordinates and variables for events (`event_*`), groups (`group_*`), and flashes (`flash_*`). An event is the signal detected by GLM over a 2 ms integration period. The group consists of contiguous events detected in adjacent sensor pixels during the 2 ms integration period. A flash is both temporally and spatially contiguous group. So, there are different numbers of each because of the parent–child relationship between the variables. Events, groups, and flashes all show total energy (`*_energy`), which is in terms of the flash radiant energy. Measures of area (`*_area`) and count (`*_count`) are useful group and flash data, which are aggregated fields.

Let's import the latitude, longitude, and event energy and combine them into a DataFrame.

```
glmLon = file_id_glm.variables['event_lon'][:]
glmLat = file_id_glm.variables['event_lat'][:]
area = file_id_glm.variables['event_energy'][:]

glmDF = pd.DataFrame({'lat': glmLat, 'lon': glmLon,
'area': area })
```

Using `scatter`, I will overlay GLM data on the Hurricane Michael plot in the previous example. The plot projection is still geostationary. However, I need to transform the GLM data from its native latitude and longitude coordinates (Plate Carrée) to the geostationary map projection. I changed the figure size and include size, color, and marker shape options for aesthetics.

```
crs = ccrs.Geostationary(central_longitude=central_lon,
    satellite_height=sat_height, globe=globe)

from_proj = ccrs.PlateCarree()
plt.figure(figsize=[20,20])

ax = plt.subplot(projection=crs)
ax.coastlines('10m', color='orange', linewidth=2)

# Plot ABI
ax.imshow(rad, origin='upper',cmap='gray_r',
extent=imgExtent)

# Add GLM data
plt.scatter(glmDF.lon, glmDF.lat, c=glmDF.area,
    transform=from_proj, s=300, marker='x')
plt.colorbar(extend='both')

plt.show()
```

From the image above, there is lightning activity close to the eyewall. Looking at the imagery, you can see there are convective towers protruding from the hurricane, indicating growth. In Section 9.3.2, I'll explore how geostationary channels can be combined to see where intensification is taking place.

7.5. Creating Maps from Datasets Using OpenDAP

I described how to import OPeNDAP data in Section 5.2.2. One of the advantages of OPeNDAP is the ease of accessing the data, rather than order data through a portal (Appendix E) you can access the data directly and immediately using a url. Xarray has OPeNDAP support built in, so I will use xarray rather than the netcdf4 package.

NOAA Extended Reconstructed (ER) Sea Surface Temperature (SST) V5 description is a global monthly sea surface temperature dataset derived from the International Comprehensive Ocean–Atmosphere Dataset (ICOADS). Production of the ERSST is on a 2° × 2° grid with spatial completeness enhanced using statistical methods. This monthly analysis begins in January 1854 continuing to the present. Since this dataset is available in the NOAA's catalog, the monthly SST can be directly imported:

```
baseURL = 'http://www.esrl.noaa.gov'
catalogURL = '/psd/thredds/dodsC/Datasets/noaa.ersst.v5/
sst.mnmean.nc'
sstID = xr.open_dataset(baseURL + catalogURL)
print(sstID)
```

```
<xarray.Dataset>
Dimensions:     (lat: 89, lon: 180, nbnds: 2, time: 1999)
Coordinates:
  * lat (lat) float32 88.0 86.0 84.0 82.0 ... -82.0 -84.0
-86.0 -88.0
  * lon (lon) float32 0.0 2.0 4.0 6.0 8.0 ... 352.0 354.0
356.0 358.0
  * time (time) datetime64[ns] 1854-01-01 ... 2020-07-01
Dimensions without coordinates: nbnds
Data variables:
    time_bnds   (time, nbnds) float64 ...
    sst         (time, lat, lon) float32 ...
```

When I import data using xarray, rather than importing a single variable, I am importing the entire dataset. The dataset structure is based on the design of Pandas' DataFrame. Using `print(sstID)`, I see that `sst` has the latitude (`lat`) and longitude (`lon`) coordinates. This dataset also has a time dimension, which begins in the year 1854 and ends in the year 2020. From the dimensions on the top line of the output, there are 1,999 time steps. Note that since this dataset is ongoing, the end date and dimensions will change to the most present value when you import it.

Below, I make a contour plot of July 2019. To get the index value, I could hard code the index 1999 (since the dates are increasing) to get the last value. I could also determine the length of the times by using `len` and by extracting the values from the data array using the values command. I subtract 1 because Python indexing starts at zero.

```
sst = sstID.sst
```

Next, I can use the `isel` command in xarray to select the time dimension value that I want to pull. The latitude and longitude coordinates do not change with time, so I do not need to subset them. The number provided to `isel` must be an index position in the array, not a specific value date of the time we are seeking.

```
mostRecent = len(sst.time.values)-1
recentSST = sst.isel(time=mostRecent)
```

The valid range of the SST goes from –1.8 to 45°C. However, below I will explicitly define the range from 0 to 30:

```
sstmin = 0
sstmax = 30
```

Finally, I can plot the data. You can extract the latitude, longitude, and SST variables directly using `contourf`. Alternatively, I use xarray `.plot` function on `recentSST`, which by default calls `pcolormesh` from matlotlib. I can also include the same `pcolormesh` options, such as `vmin=` and `vmax=`. Xarray's plot will automatically pass the latitude and longitude coordinates to `pcolormesh`. Thus, I do not need to import the latitude and longitude, as long as these coordinates are linked to the `sst` variable in the netCDF file.

```
fig = plt.figure(figsize=[10,5])
ax = plt.subplot(projection=ccrs.Orthographic(-90, 0))
recentSST.plot.contourf(levels=levels, cmap=plt.get_cmap
('plasma'), transform=ccrs.PlateCarree())
ax.coastlines('50m')

plt.show()
```

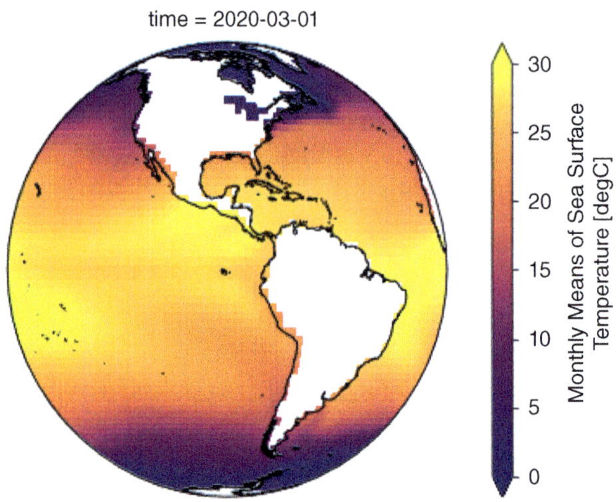

I can also create a contour plot by changing the code above from `.plot(...)` to `.plot.contourf(...)`. To customize the contour levels, I use the `levels=` argument (rather than `vmin=` and `vmax=`), just as I would for contourf in matplotlib. I can use np.arange to increment the data range that I defined earlier (0 to 30) by 2, to explicitly define where I want to see contour lines.

```
levels = np.arange(sstmin, sstmax, 2)

fig = plt.figure(figsize=[10,5])
ax = plt.subplot(projection=ccrs.Orthographic(-90, 0))
recentSST.plot.contourf(levels=levels, cmap=plt.get_cmap
('plasma'),
    transform=ccrs.PlateCarree())
ax.coastlines('50m')
plt.show()
```

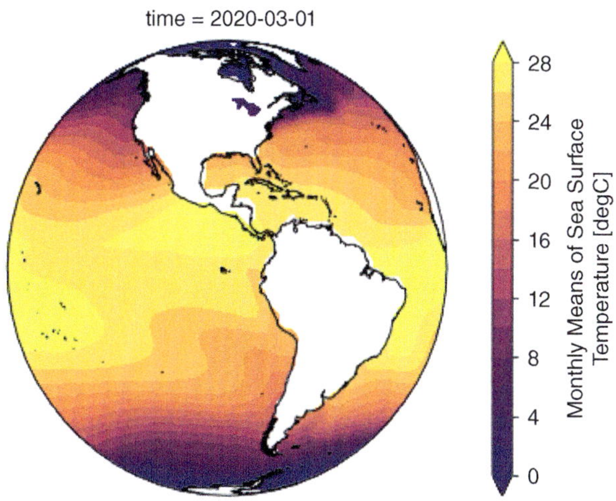

7.6. Summary

In this chapter, I discussed how to take two- and three-dimensional plots and project them onto a variety of maps. Using Cartopy, I showed how you can create figures by combining the same Matplotlib steps you learned in the previous chapter with overlays of country boundaries or other geographic features. To change from one coordinate system to another, the axes and the data must both be labeled with their respective projection. While Cartopy package does much of the mathematical work for us, it is helpful to understand the differences between the projections to choose the best one for your project. So far, I have worked with the data in its native grid. In the next chapter, I will show examples of how to change the dimensionality and gridding of the data itself. This will be useful for comparing and combining datasets.

References

CGMS (2013, October 30). LRIT/HRIT Global Specification. Retrieved from https://www.cgms-info.org/documents/cgms-lrit-hrit-global-specification-(v2-8-of-30-oct-2013).pdf

Ordnance Survey. (2018). A Guide to Coordinate Systems in Great Britain. Retrieved from https://www.ordnancesurvey.co.uk/documents/resources/guide-coordinate-systems-great-britain.pdf

Snyder, John Parr (1987). Map projections – a working manual. U.S. Government Printing Office. p. 192.

U.S. Department of the Interior U.S. Geological Survey. Map projections. Retrieved from https://store.usgs.gov/assets/mod/storefiles/PDF/16573.pdf

8

GRIDDING OPERATIONS

Earth scientists often combine data, either from point observations or from other satellites, to perform complex analysis and advanced visualizations. For a single point of data, users must locate the nearest pixel in time and space to a point on the ground to obtain a unique match. For multiple datasets, one or both datasets may need to be converted to an entirely new grid for them to be combined. This chapter contains tutorials that show some basic gridding operations within Python to convert between grids. Grid changes can be performed manually using user-defined functions or with useful packages like scipy and pyresample. Additionally, satellite data are not always stored in a regularly spaced latitude–longitude grid (e.g., every 1°) so it is useful to consider how to work with data that are spaced irregularly, such as across swaths of data.

As you may have observed in Chapter 7, swath data at the scan edges can be distorted because the signal at the far end of the beam travels a further distance than the edge that is closer. This effect is further distorted by the curvature of the Earth, particularly at high latitudes. Data developers thus have a choice when designing the structure of the datasets: leave the data in its native form or converted it to a new grid. There is no standard grid in remote sensing datasets. Instead, the decision is often driven by end-user needs or by constraints in the retrieval algorithm. For instance, daily average AOD (a Level 3 product) is gridded to 1.0 × 1.0 degree whereas VIIRS-based JPSS Level 2 datasets are mainly stored in their native swaths. The grid can also be motivated by the overall file size of the dataset. Some of the GOES-R datasets, such as the shortwave radiation budget, are stored into larger 50 km grids even though it is possible to

Earth Observation Using Python: A Practical Programming Guide, Special Publications 75, First Edition. Rebekah B. Esmaili.

DOI: 10.1002/9781119606925.ch8

perform a retrieval at finer scales. When using satellite data, you may want to change the resolution or gridding of the dataset. This can make it easier (and faster) to plot the data using the methods you've seen so far, or perhaps you may wish to combine and compare multiple datasets, which you will learn more about in the next chapter.

This chapter uses the following packages that you have already learned about: **NumPy, Pandas, netCDF4, h5py, matplotlib.pyplot,** and **Cartopy.** The examples will show how to manually do simpler grid conversions from one grid to another, as well as apply some advanced operations using package functions. The examples in this chapter will discuss and use the following packages:

- **scipy.interpolate**
- **pyresample**
- **matplotlib.patches**

To learn about gridding operations, it is useful to gain experience working with a variety of different input grid formats. I will also use the following datasets:

- **VIIRSNDE_global2018312.v1.0.txt.** This dataset is formatted as a list of values, which you will convert to a two-dimensional grid.
- **3B-HHR.MS.MRG.3IMERG.20160811-S233000-E235959.1410.V06B. HDF5.** This hdf dataset has data structures in one-dimensional regular spacing, which you will convert to two dimensions.
- **OR_ABI-L1b-RadM1-M3C02_G16_s20182822019282_e20182822019339_c20182822019374.nc.** A GOES-16 ABI mesoscale image from the red visible band of Hurricane Michael from 9 October 2018. The data have a resolution of 0.5 km, so I will show how you can quickly resample the data to 10 km using NumPy indexing.
- **OR_ABI-L1b-RadM1-M3C13_G16_s20182822019282_e20182822019350_c20182822019384.nc.** This dataset is also from the ABI and at same time as the above dataset, but uses a longwave infrared band (2 km resolution). You will use the `scipy.interpolate` to explore how different interpolation methods change the final image.
- **JRR-AOD_v1r1_npp_s201808091955538_e201808091957180_ c201808092049460.nc.** The example here will use **pyresample** to quickly convert from swath data to a linearized grid.

8.1. Regular One-Dimensional Grids

I showed examples of how to bin (aggregate into groups) data when I discussed histograms and bar plots in Section 6.1. Similarly, data can be spatially aggregated, and therefore changing the grid, to show frequency by location. The following example uses the one-dimensional data from the VIIRS fire product dataset. The scatter plot of fire locations in Section 7.2.1 is useful for seeing their global distribution. You may notice that the fires are often clustered close together, so it can be helpful to count the fires in a 1.0 degree grid and plot using `pcolormesh` from matplotlib. Below, I import the libraries I need for this chapter. These packages have all been covered in past chapters, aside from **scipy.interpolate**.

```
import pandas as pd
import numpy as np
from netCDF4 import Dataset
import h5py
from matplotlib import pyplot as plt
from cartopy import crs as ccrs
import scipy.interpolate
```

Next, I will import the VIIRS fire product dataset using Pandas and use `head()` to inspect the first few rows.

```
fires = pd.read_csv("data/VIIRSNDE_global2018312.v1.0.
txt")
fires.head()
```

	Lon	Lat	Mask	Conf	brt_t13(K)	frp(MW)	line	sample	Sat	YearDay	Hour
0	27.110006	30.769241	8	52	302.877533	5.814295	242	1735	NDE	2018312	1
1	26.083252	30.534357	9	100	332.959717	24.340988	301	1620	NDE	2018312	1
2	34.865997	28.162659	8	38	301.165985	6.107953	396	2589	NDE	2018312	1
3	34.872623	28.161121	8	71	307.277985	9.287819	396	2590	NDE	2018312	1
4	34.865070	28.158880	8	39	301.227783	6.001442	402	2590	NDE	2018312	1

In the `fires` dataset, the latitude and longitude columns (`LAT` and `LON`) have values that extend out six decimal places. Thus, it will be difficult to group these values unless I round these columns. So, for this example, I will create a new grid that has 1.0 degree spacing. Recall that `pcolormesh` requires a two-dimensional `meshgrid` input for latitude, longitude, and the plotting variable whereas the `fires` is a one-dimensional list. So, I need to restructure the data in addition to binning it. The command `np.mgrid` will return two variables each with a two-dimensional grid using the following syntax:

```
np.mgrid[ start_lon:end_lon: num_points_x,
start_lat:end_lat: num_points_y ]
```

So, I need to define these six parameters. First, I define the latitude and longitude `coverage`, which spans –90.0 (`start_lat`) to 90.0 (`end_lat`) and –180.0 (`start_lon`) to 180.0 (`end_lon`) in this dataset. Note that some datasets define longitude from 0 to 360 or use letters rather than the negative sign to indicate north, south, east, or west. To determine the number of points (num_points_x, num_-points_y) of this new dataset along the latitude and longitude coordinates, I can subtract the endpoints (e.g., coverage[2] - coverage[0] for longitude) and divide by the grid_size, which I equate to 1.0 to represent the 1.0 degree spacing (for practice, you can pick another resolution and go through the same steps). Since the number of points cannot be a fraction, I use int() to round down to the nearest whole value integer. In some languages, rounding down to the nearest integer is called **flooring**, versus **ceiling,** which rounds up to the nearest integer. When I print num_points_x and num_points_y, you can see that there are 360 and 180 longitude and latitude points, respectively.

```
# Number of nx and ny points for the grid. 720 nx, 360 ny
creates 1.0 degree grid.
coverage = [-180.0 , -90.0 , 180.0 , 90.0]

grid_size = 1.0

num_points_x = int((coverage[2] - coverage[0])/grid_size)
num_points_y = int((coverage[3] - coverage[1])/grid_size)

print(num_points_x, num_points_y)
360 180
```

I now have all the arguments for mgrid. Note that I need to pass these arguments as **complex numbers**, as in, a real + imaginary number (imaginary denoted by j in python). If I pass a complex number into mgrid, the resulting list will include the final value in the series (**inclusive behavior**). If I pass real values into mgrid, the last value is not added to the array (**exclusive behavior**). So, I need to pass end_lon + grid_size in order to not crop the final value. Alternatively, as mentioned, I can pass num_points_x and num_points_y in complex form using complex(). A simplified example of inclusive versus excusive behavior follows:

```
# Using a real step length will skip the end value
(exclusive)
print(np.mgrid[0:4:1.0])

# Using a complex number of bins will include the end value
(inclusive)
print(np.mgrid[0:4:5j])
[ 0.  1.  2.  3.]
[ 0.  1.  2.  3.  4.]
```

Neither method is right or wrong, but I prefer the simplicity of the complex approach. Now, I can compute the two-dimensional longitude (Xnew) and latitude (Ynew) grid:

```
nx = complex(0, num_points_lon)
ny = complex(0, num_points_lat)

Xnew, Ynew = np.mgrid[coverage[0]:coverage[2]:nx,
coverage[1]:coverage[3]:ny]
```

Now that the coordinate grid is defined, I need to fill a third array with the fire counts. I create an array (fire_count) with zeros that is the same size and Xnew and Ynew. Then, I populate fire_count with the number of fires within the 1.0 degree latitude and longitude bin using a for loop.

The for loop below iterates over the fires.Lon column. The enumerate function returns the index (i), from which I can get the fires.Lat value. For each row, the code below calculates the index in the new grid (latbin and lonbin) and increases the fire count by 1. Note that loops in Python can be very slow for large datasets. If you are working with two-dimensional data, for instance, it will be better to use griddata, which I discuss in the next section.

```
fire_count = np.zeros([num_points_x, num_points_y])

for i, lon in enumerate(fires.Lon):
    lat = fires.Lat[i]

    adjlat = ((lat + 90) / grid_size);
    adjlon = ((lon + 180 ) / grid_size);

    latbin = int(adjlat)
    lonbin = int(adjlon)

    fire_count[lonbin, latbin] = fire_count[lonbin, latbin]+1
```

Since fire_count contains more zeros than values, if I were to pass the values as-is into pcolormesh, most of the plot will be filled with zeros. It is easier to visualize the fire by replacing the zeros in fire_count with NaN values (np.nan) so that pcolormesh will not display these values.

```
fire_count[fire_count == 0] = np.nan
```

To create the figures, I use subplots (discussed in Section 6.3.6) so I can compare the original scatter plot to the binned density plot I created in the steps above. I can use the subplot_kw option to pass the projection to all the plots within the figure. Note that if you are comparing figures with different projections, you will need to use the subplot (nonplural) command to specify the projection of plot individually.

The original data (`fires`) is one-dimensional, so I use `scatter` in the first plot. In the second plot, `fires_count` is two-dimensional so I use `pcolormesh`.

```
fig, (ax1, ax2) = plt.subplots(nrows=2, figsize=[15, 15], \
    subplot_kw={'projection': ccrs.PlateCarree()})

ax1.coastlines()
ax1.set_global()
ax1.scatter(fires['Lon'], fires['Lat'], s=1)

ax2.coastlines()
ax2.set_global()

ax2.pcolormesh(Xnew, Ynew, fire_count, vmin=0, vmax=40)

plt.show()
```

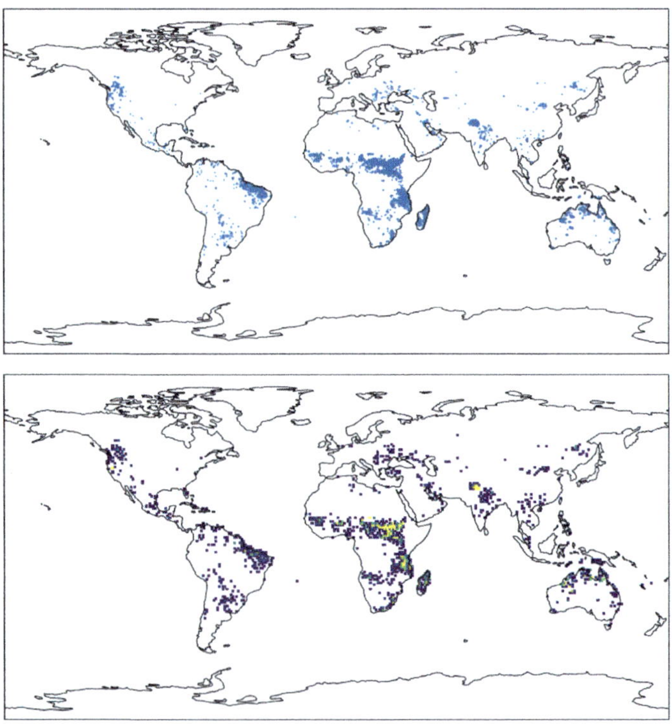

Compared to the scatter plot, it is easier to see where a large number of fire hotspots are distributed globally after binning into a larger grid. The steps above are useful for working with smaller, one-dimensional data such as DataFrames. In the next section, I discuss how two-dimensional data with both regular and irregular grids can be aggregated.

8.2. Regular Two-Dimensional Grids

In the previous section, the `fires` DataFrame contained 11,833 rows of daily observations, which is a relatively small dataset. In the next example, I will change spatial grid of precipitation from IMERG, which has 6,480,000 observations every half hour. Since `for` loops can become slow in Python, they should be avoided for larger datasets. Instead, I recommend using the `griddata` function to perform a two-dimensional interpolation from one grid to another.

I begin by importing the HDF data (Section 5.1.3). I will need the precipitation rate and coordinates. In the HDF file, the first dimension of precipitation refers to the time dimension, but there is only one element because each file represents one snapshot of data, so I will simplify the array by making it two-dimensional instead of three.

```
fname = 'data/3B-HHR.MS.MRG.3IMERG.20160811-S233000-
E235959.1410.V06B.HDF5'
imergv6 = h5py.File(fname, 'r')

precip = imergv6['Grid/precipitationCal'][0,:,:]
lat = imergv6['Grid/lat'][:]
lon = imergv6['Grid/lon'][:]
```

As you can see from above (or if you inspected the file from beforehand), the latitude and longitude coordinates are one-dimensional while precipitation is two-dimensional. This is because the data are linear and regularly gridded across latitude and longitude coordinates; that is, the coordinates are spaced in 0.1 degree increments. Other regularly spaced data could be incremented by distance (e.g., 4 km) or linearized after applying a rescaling function. Data can also be regularly spaced in the vertical dimension as well.

In the IMERG dataset, the coordinates are stored in one dimension, which helps to keep the file size small. The next example shows how to create a two-dimensional mesh using `np.meshgrid`.

You can check for yourself using `lon.shape` and `lat.shape`, but the dimension of the `lat` variable is 1800 while the `lon` variable is 3600. The default dimension of the two output arrays will be (1800, 3600) using the `indexing='xy'` option. If you set `indexing='ij'`, the resulting output will be

(3600, 1800), the transpose of the original array. In the remainder of the example, the array dimensions are organized in (3600, 1800) order and I will use the `indexing='ij'` keyword to keep the dimensions consistent.

```
Xold, Yold = np.meshgrid(lon, lat, indexing='ij')
```

Now that the old grid is defined, it is time to define the new grid. In the next example, I will show how to write a function (`create_2d_grid`) to create a two-dimensional grid using `np.mgrid`, which I discussed in the previous section. The following code is very similar to the example already shown. I recommend writing common tasks as functions because the code is easier to reuse in other parts of your notebook or in other programs. One difference is that the following function is written so that it takes the input argument `grid_size`, which is the grid spacing in degrees. Now we can change the spacing dynamically, without rewriting the code.

```
def create_2d_grid(grid_size):
    coverage = [-180.0 , -90.0 , 180.0 , 90.0]

    num_points_x = int((coverage[2] - coverage[0])/
grid_size)
    num_points_y = int((coverage[3] - coverage[1])/
grid_size)

    nx = complex(0, num_points_x)
    ny = complex(0, num_points_y)

    Xnew, Ynew = np.mgrid[coverage[0]:coverage[2]:nx,
coverage[1]:coverage[3]:ny]
    return Xnew, Ynew
```

Now call the `create_2d_grid` function and pass in a grid size of 0.5. If you are curious, you can try another value.

```
Xnew, Ynew = create_2d_grid(0.5)
```

To check if the function returns the expected value, examine the size of the output variables. The `precip`, `Xold`, and `Yold` variables should have the same dimensions, and the new grid should be smaller since the resolution is being reduced from 0.1 to 0.5°:

```
precip.shape, Yold.shape, Ynew.shape
((3600, 1800), (3600, 1800), (720, 360))
```

Since I want to interpolate my grid in two dimensions, I will use the griddata function from scipy.interpolate. You can read more about this function by typing ?scipy.interpolate.griddata. Here is the basic syntax for griddata:

```
scipy.interpolate.griddata(old_points, old_values,
new_points)
```

I have all the required input arguments: the coordinates and values from the original grid (old_points, new_points), as well as our new coordinate grid (new_points). However, the data must be reformatted in a form that griddata can interpret. The argument for old_points needs to be in an (x,y) tuple form, where x and y are in one-dimensional. The second argument, old_values, needs to be one-dimensional as well. The third argument, new_points, is a two-dimensional tuple of the form (xnew, ynew), where xnew and ynew are both two-dimensional.

In the code block below, I use NumPy's flatten() to condense our two-dimensional precip variable into a one-dimensional array, which I save in the values variable. Then, I create an empty array (points) with the dimensions values.shape[0] and 2 (Alternatively, I could have also used np.size (precip) instead of values.shape[0]). Again, using flatten(), I save the one-dimensional x-values in the first column and the flattened y-values to the second:

```
values = precip.flatten()

dims = (values.shape[0], 2)
points = np.zeros(dims)

points[:, 0] = Xold.flatten()
points[:, 1] = Yold.flatten()
```

The next step may be a little slow, depending on your resolution choice and computer's memory and processor. If I performed this computation manually, I would be writing over 6 million points to 260,000, for a total of 1,679,616,000,000 computations. The griddata function is optimized to perform better than this, but keep in mind that this can be a computationally expensive task. Decreasing the resolution of the final grid can speed the process up.

Below, I use nearest neighbor interpolation, but linear and cubic are also options, examined further in Section 8.3.2. Your choice in interpolation scheme will differ, depending on field and applications. From my personal experience, atmospheric remote sensing fields tend to prefer displaying the nearest neighbor, whereas modelers, forecasters, and planners prefer additional smoothing and may use linear and cubic methods. Feel free to experiment with the output below using different techniques. In terms of speed, nearest neighbor is the fastest, while cubic is the slowest of the three choices.

```
gridOut = scipy.interpolate.griddata(points, values,
(Xnew, Ynew),
method='nearest')
```

Finally, I can create the plot showing the IMERG data on the lower resolution grid. I want to compare the original grid to the new grid that I defined, so I will display these two plots side by side (Section 6.2.6). To simplify the code, I wrote a short loop to iterate over the axes (`for ax in axes`) to set the coastlines and extent for both subplots. This is followed by adding the original grid and the lower resolution grid, using `pcolormesh`:

```
fig, axes = plt.subplots(ncols=2, figsize=[15, 15],
   subplot_kw={'projection': ccrs.PlateCarree()})

for ax in axes:
    ax.set_extent([-94.3, -88.8, 28.9, 33.1])
    ax.coastlines('10m', color='orange')

axes[0].set_title("Before regridding")
axes[0].pcolormesh(Xold, Yold, precip, vmin=0, vmax=20)

axes[1].set_title("After regridding")
axes[1].pcolormesh(Xnew, Ynew, gridOut, vmin=0, vmax=20)

plt.show()
```

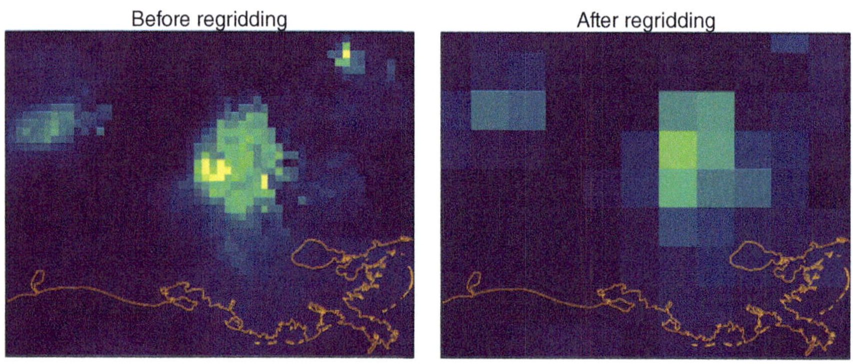

So, in the previous example, the resolution is lowered (a process sometimes called **coarsening** or **upscaling**) for regularly spaced data. The `griddata` function does not require that the original grid be linearly spaced, but if it is not, the computation time can dramatically increase. In the next section, I will address working with irregular grids which also often result in very large datasets.

8.3. Irregular Two-Dimensional Grids

8.3.1. Resizing

Gridding can be computationally expensive for very high-resolution datasets. There are cases where you do not need a rigorous sampling method to change the data's resolution. For example, you may want to just visually inspect two data products and not perform a thorough quantitative analysis. You can quickly resize data using NumPy's iteration techniques. Note that the following resizing method produces significant sampling errors for satellite data, particularly at the scan edges, so it is not appropriate for quantitative analysis. However, it is helpful for "quick looks" of datasets.

The visible and IR channels on imagers tend not to have the same resolution, which I will discuss more in Chapter 9. Below, I import the 0.47 μm radiance (`Rad`) from an ABI mesoscale image of Hurricane Michael.

```
fname = "data/goes-meso/michael/OR_ABI-L1b-RadM1-
M3C02_G16_s20182822019282_e20182822019339_
c20182822019374.nc"
g16nc = Dataset(fname, 'r')
C02 = g16nc.variables['Rad'][:,:]
```

This example follows the same steps from Section 7.4 to create a geostationary grid for the GOES-16 ABI, so refer to that section if the following summary moves too quickly. First, I pull metadata about the ABI projection, such as the satellite height (`sat_height`), the major and minor axes (`semi_major` and `semi_minor`), and central longitude (`central_lon`). The variables are used to define an object that I called `globe`. I then import the `x` and `y` variables from the dataset and multiply them times the satellite height. I define the `imgExtent` variable using the minimum and maximum coordinates because it will be useful later when I call matplotlib and make the plot. In the last line, `globe` is passed into the Cartopy geostationary projection (`ccrs.Geostationary`) along with the central longitude and height to define the geostationary grid.

```
proj_var = g16nc.variables['goes_imager_projection']

sat_height = proj_var.perspective_point_height
semi_major = proj_var.semi_major_axis
semi_minor = proj_var.semi_minor_axis
central_lon = proj_var.longitude_of_projection_origin
globe = ccrs.Globe(semimajor_axis=semi_major,
semiminor_axis=semi_minor)

# Multiply the x, y coordinated by satellite height to get
the pixel position
X = g16nc.variables['x'][:] * sat_height
Y = g16nc.variables['y'][:] * sat_height
imgExtent = (X.min(), X.max(), Y.min(), Y.max())

crs = ccrs.Geostationary(central_longitude=central_ lon,
satellite_height=sat_height, globe=globe)
```

Now that the data are imported, let's say for example I want to resize the data so that it is 20x smaller in each dimension (400x smaller in terms of area).

```
resized_C02 = C02[::20, ::20]
```

Subplots can be used to compare how this data looks in its original scale with the new reduced resolution. Aside from adding `fig.colorbar` and using `imshow` in place of `pcolormesh`, the plotting code below is like the previous example:

```
lims = [0, 500]
fig, axes = plt.subplots(ncols=2,  figsize=[15, 15],
subplot_kw={'projection': crs})

for ax in axes:
    ax.set_extent([-90.0, -82.0, 22.0, 30.0])
    ax.coastlines('10m', color='black')

axes[0].set_title("Native Resolution")
im1 = axes[0].imshow(C02, cmap=plt.get_cmap("rainbow"),
extent=imgExtent, vmin=lims[0], vmax=lims[1],origin='upper')

axes[1].set_title("20x Lower Resolution")
im2= axes[1].imshow(resized_C02, cmap=plt.get_cmap("rainbow"),
extent=imgExtent, vmin=lims[0], vmax=lims[1]origin='upper')
```

```
# Adding colorbars
fig.colorbar(im1, pad=0.05, orientation='horizontal',
ax=axes[0])
fig.colorbar(im2, pad=0.05, orientation='horizontal',
ax=axes[1])

plt.show()
```

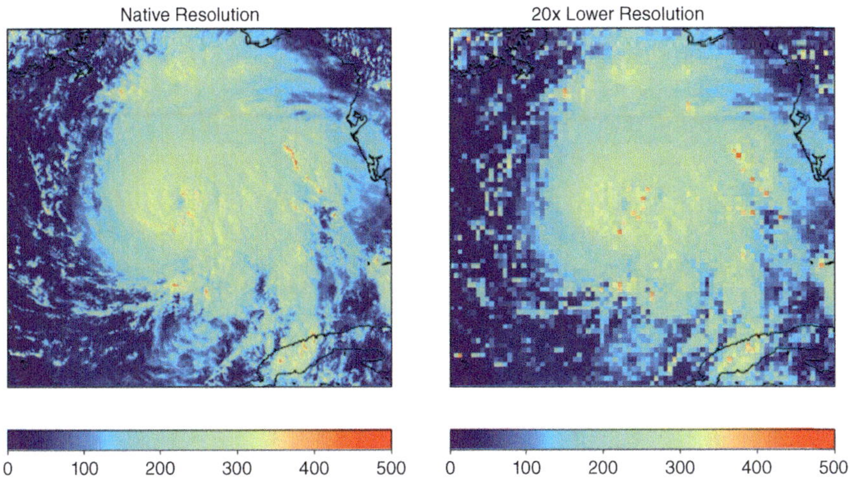

With a 20x reduction in resolution, the data are noticeably coarser and illustrate some of the errors that can arise from this method. Smaller resolution reductions will not be as noticeable. For instance, if you repeat this example but instead use `resized_C02 = C02[::2, ::2]` the differences would be difficult to detect.

The method does not perform any type of averaging and is not appropriate for quantitative comparisons. However, it is useful for quick looks of the data or if you are simply using the image as a background for other datasets.

8.3.2. Regridding

Earlier, I discussed how `griddata` (part of the **scipy** package) can regrid regular data. `Griddata` can also be used to match the increase the resolution of one dataset to match another. Let's import GOES-16 Channel 13 (10.3 μm):

```
fname = "data/goes-meso/michael/OR_ABI-L1b-RadM1-
M3C13_G16_s20182822019282_e20182822019350_
c20182822019384.nc"
goesnc_sm = Dataset(fname, 'r')
C13 = goesnc_sm.variables['Rad'][:]
sm_x = goesnc_sm.variables['x'][:]
sm_y = goesnc_sm.variables['y'][:]
```

In the previous example, the GOES-16 Channel 2 dimensions (using `y.shape` or `x.shape`) were 2000 × 2000. In contrast, Channel 13 is 500 × 500 (using `x_sm.shape` or `y_sm.shape`). In this example, I will increase the resolution of Channel 13 to match Channel 2. It will be easiest to see the differences if I zoom into a spot on the image. First, I introduce some code to annotate the plot to define the region that that I will zooming into. The `patches` function inside `matplotlib.patches` can be used to help draw a box on the maps:

```
import matplotlib.patches as patches
```

Below, I define the region on the plot where I want to draw a box (arbitrarily chosen):

```
vmin = 0
vmax = 50

delta = 0.005
xbot = -0.025
ybot =  0.075

lower_left = (xbot, ybot)
```

For simplicity, I am omitting the Cartopy components in the plot code below. I reversed the color scale (by adding an `_r` after the colormap name) below so that the brightest part of the scale corresponds to the hurricane, not the background. To annotate the plot with the rectangle, I need to define the lower-left corner (`lower_left`), and the rectangle height and width (both using `delta`, since they are the same in this example):

```
fig, axes = plt.subplots(ncols=1, nrows=1, figsize=[8,8])

meso = axes.pcolormesh(sm_x, sm_y, C13, vmin=vmin,
vmax=vmax,
```

```
cmap=plt.get_cmap("tab20c"))
fig.colorbar(meso, label='Brightness Temperature (K)')
zoom_box = patches.Rectangle(lower_left, delta, delta,
linewidth=1, edgecolor='r', facecolor='none')
axes.add_patch(zoom_box)
axes.set_aspect('equal')

plt.show()
```

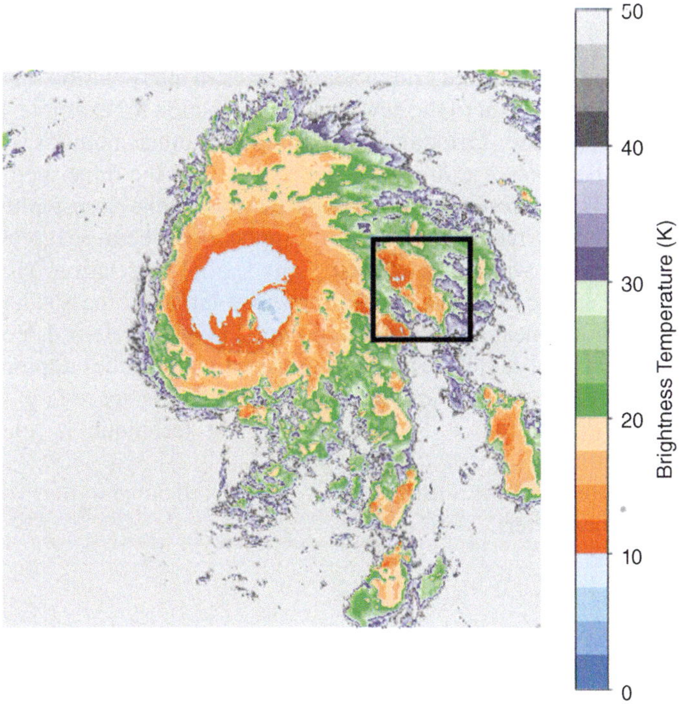

In the regular, two-dimensional grid example in Section 8.2, both the datasets existing grid (the "old" grid) and the one I want to map to (the "new" grid) must be defined. In this example, the old grid is lower-resolution Channel 13 grid and the new grid is the higher-resolution Channel 2 grid. Since the data in the netCDF file are one-dimensional, I will use `meshgrid` (Section 6.4.1) to transform it to two-dimensional:

```
Xold, Yold = np.meshgrid(sm_x, sm_y)
Xnew, Ynew = np.meshgrid(x, y)
```

Following the steps from Section 8.2, `griddata` requires inputs to be a tuple of one-dimensional x and y points and a one-dimensional array of values. Below, I flatten the channel 13 radiance data (`C13`) and create an empty array, which I fill with the flattened coordinates:

```
values = C13.flatten()

dims = (values.shape[0], 2)
points = np.zeros(dims)

points[:, 0] = Xold.flatten()
points[:, 1] = Yold.flatten()
```

When moving from a small grid to a larger one, an interpolation scheme must be selected to fit the old data to the new grid. In the Section 8.2 example, I used the nearest neighbor method. Griddata can resample in other methods: linear and cubic. Figure 8.1 shows an example of the impact that the three methods have on the new grid. The nearest neighbor repeats the value of the closest grid cell. This approach is often preferred in remote sensing applications, because it preserves the original measurement without introducing any spatial correlation. However, if satellite data are sparse, repeating edge values over large distances can introduce errors. In such cases, linear and cubic interpolation may be preferred. You can see below that the resulting data have a smoother transition from the endpoints. While not shown, in some cases the variable can be linear in log space (e.g., trace gas concentrations with altitude). So, the appropriate technique is going to be application specific.

I will compute `griddata` using nearest, linear, and cubic so that you can see how these methods differ.

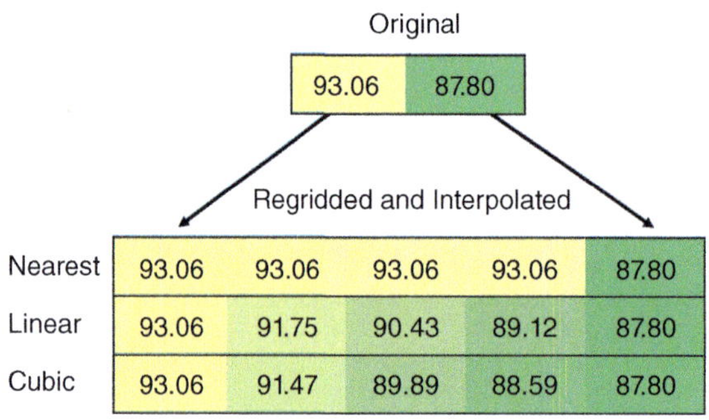

Figure 8.1 An example of differences in the nearest neighbor, linear, and cubic interpolation method on the values used in the gridded data.

```
gridOut_nn = scipy.interpolate.griddata(points, values,
(Xnew, Ynew),
method='nearest')
gridOut_lin = scipy.interpolate.griddata(points, values,
(Xnew, Ynew),
method='linear')
gridOut_cube = scipy.interpolate.griddata(points, values,
(Xnew, Ynew),
method='cubic')
```

For completeness, also look at the re-binning method that I used in Section 8.3.1:

```
gridOut_rebin = np.repeat(np.repeat(C13, 4, axis=1), 4,
axis=0)
```

I make a 2 × 2 plot with all these methods side by side (using `fig.subplots`) to compare the differences. I could write the code for a single plot, and then copy and paste the same code four times and change a z variable to the gridded one (`gridOut_rebin`, `gridOut_nn`, `GridOut_lin`, `GridOut_cube`). Instead, I will write a `for` loop to pass the arguments into the subplot to shorten and simplify my code. Below I define a list nested in another list with each of our 2 × 2 axes. In the same manner, I add plot labels so that I can differentiate the contents of each subplot:

```
gridOut = [[gridOut_rebin, gridOut_nn], [gridOut_lin,
gridOut_cube]]
labels = [["Re-binning", "Nearest Neighbor"],["Linear",
"Cubic"]]
plt.rcParams.update({'font.size': 18, 'figure.figsize':
[20, 20]})
```

Finally, I can create my plot. Since this plot contains four subplots, the loops help shorten the code substantially:

```
fig, axes = plt.subplots(ncols=2, nrows=2)

cmap=plt.get_cmap("viridis_r")

for i, axis in enumerate(axes):
    for j, ax in enumerate(axis):
        ax.set_xlim([xbot, xbot+delta])
        ax.set_ylim([ybot, ybot+delta])
```

```
        ax.set_aspect('equal')
        ax.set_title(labels[i][j])
        ax.pcolormesh(Xnew, Ynew, gridOut[i][j], vmin=vmin,
vmax=vmax, cmap=cmap)

plt.show()
```

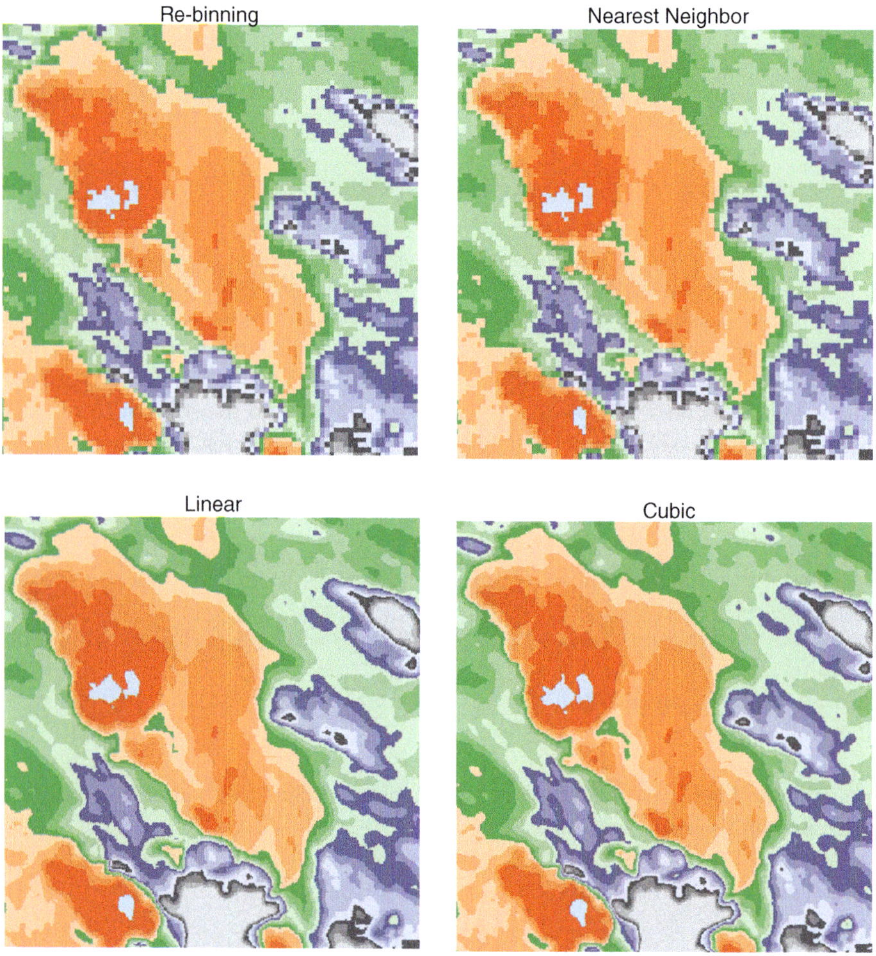

The upper two look more similar: in both the re-binning and nearest neighbor methods, the values are not being altered; the difference is which is being selected to fill the expanded grid. Meanwhile, the linear and cubic methods both produce smoother final images. As mentioned, there is no "correct" method; the one you ultimately decide to use will be application specific. For "quick looks," re-binning is often enough and requires fewer lines of code. However, for more thorough analysis, I recommend using one of the `griddata` methods. If speed is your main concern, the average run time for this example on my personal computer is as follows: re-binning (13.2 ms), nearest (5.2 s), linear (5.9 s), and cubic (10.7 s). Note that the exact runtimes will vary from computer to computer.

8.3.3. Resampling

If you need a robust gridding technique while working with a voluminous dataset, then the resizing technique will produce too many errors and `griddata` may be too slow. Below, I will employ the **pyresample** package to efficiently resample the data. Like Cartopy and netCDF, pyresample is not included with the Anaconda distribution, so you will need to install it before proceeding.

```
from pyresample import geometry
from pyresample.kd_tree import resample_nearest
```

I will use the JPSS AOD product in the following example, since it is both high resolution and on an irregular grid. First, I import the relevant fields from the netCDF file:

```
fname='data/JRR-AOD_v1r1_npp_s201808091955538_
e201808091957180_ c201808092049460.nc'
file_id_NPP = Dataset(fname)

aod = file_id_NPP.variables['AOD550'][:,:]
lat = file_id_NPP.variables['Latitude'][:,:]
lon = file_id_NPP.variables['Longitude'][:,:]
```

Like in the previous examples, I will create our native grid and coordinates, as well as the new grid. The coordinates and values are stored in two dimensions in the netCDF file, so they don't need to be converted into a mesh grid.

Pyresample employs a different method of defining the grids. Since Level 2 polar orbiting data are often stored as irregularly gridded swaths in the files, I can use pyresample's `SwathDefinition` to define it. Additionally, pyresample is designed specifically for satellite observations of the Earth. It allows you to directly pass the two-dimensional latitude (`lat`) and longitude (`lon`) coordinates:

```
# Input list of swath points
oldLonLat = geometry.SwathDefinition(lons=lon, lats=lat)
```

Next, I will create a new 0.1° grid to project the irregular data to, making it regular. To save some computing time, I will only use the grid that surrounds the swath of data. If instead you wanted to process a global grid with 0.1° spacing, you can change the parameters in np.arange() to (−180, 180, 0.1) for × and (−90, 90, 0.1) for y. However, this will produce 3600 × 1800 points and may be slow on your machine. Since the new grid is only a rough estimate, below I use `np.arange` to generate the list of coordinates, which are then passed into `meshgrid`.

```
# Create a new grid at 0.1 degree resolution
x = np.arange(lon.min(), lon.max(), 0.1)
y = np.arange(lat.min(), lat.max(), 0.1)
newLon, newLat = np.meshgrid(x, y)
```

To define the new grid in pyresample, use a command named `GridDefinition`.

```
# define the new grid using
newLonLat = geometry.GridDefinition(lons=newLon,
lats=newLat)
```

Finally, use `resample_nearest` to convert `lat`, `lon`, and `aod` to the new 0.1° regular grid. The argument `radius_of_influence` (in meters) determines how far the neighbor search will look for a match. Because there are missing values in swath data, it is helpful to keep the radius small (which also reduces processing time), so I set it to 5000 m. If `fill_value=none` and no match is found, a NaN is returned, and that pixel will not plot. These last two arguments are necessary because they will exclude data outside of the swath. In comparison, `griddata` will search for a neighbor no matter how far away, which can create unrealistic effects at swath edges. If using `griddata`, you would need to create a mask to filter values that are outside of a swath.

```
# Resample the data
newAOD = resample_nearest(oldLonLat, aod, newLonLat,
radius_of_influence=5000, fill_value=None)
```

This `resample_nearest` function may have run faster than the equivalent `griddata` method. Finally, the original and new grid are compared side by side:

```
lims = [0,1]
fig, axes = plt.subplots(nrows=2,  figsize=[15, 10],
    subplot_kw={'projection': ccrs.PlateCarree()})

for ax in axes:
    ax.coastlines('10m', color='black')

axes[0].set_title("Before regridding")
```

```
axes[0].pcolormesh(lon, lat, aod, vmin=lims[0],
vmax=lims[1])

axes[1].set_title("After regridding")
axes[1].pcolormesh(newLon, newLat, newAOD, vmin=lims[0],
vmax=lims[1])

plt.show()
```

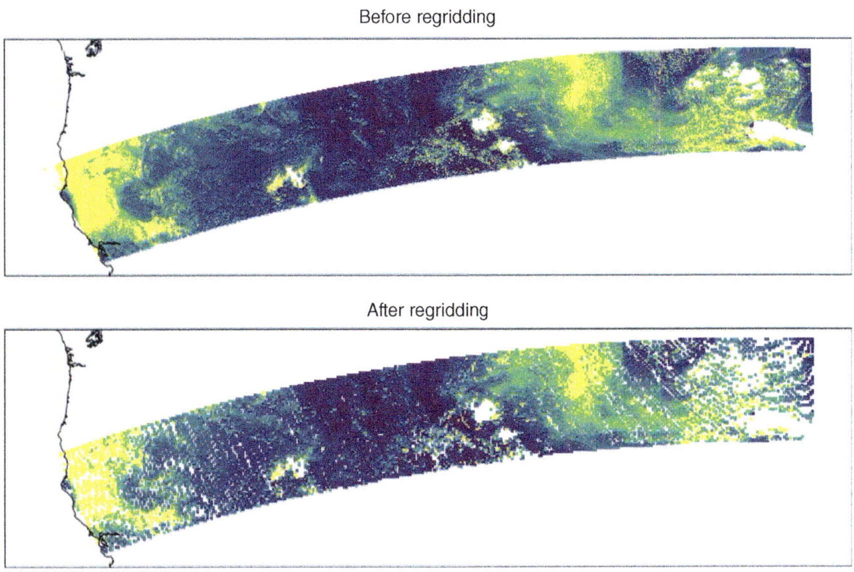

There is some distortion at the scan edges, where pixels are longer horizontally than they are tall. Adjusting the `radius_of_influence` option can help fill in these gaps, although doing so may not necessarily correlate with ground observations. In addition to `resample_nearest`, Pyresample also has other methods, such as `bilinear` and `resample_gauss`, which are worth exploring. While a relatively new package, Pyresample is optimized for satellite data and is a useful package for working with irregularly gridded data.

8.4. Summary

In this chapter, you learned how to do basic transformations between grids that are stored in a variety of formats. You aggregated irregularly spaced data to a regularly spaced dataset in one-dimensional, decreasing the resolution of both regularly and irregularly spaced two-dimensional data, and compare different spatial interpolating packages and methods. This chapter serves as an introduction, as there are many more complex gridding schemes depending on the area of research. Gridding operations are necessary to combine datasets together, either to assimilate different datasets, compare them, or to create new analyses and understanding. In the next chapter, I'll show you how datasets can be combined to study specific phenomena and features on the surface and in the atmosphere.

References

Jones, Philip W. (1999). First- and second-order conservative remapping schemes for grids in spherical coordinates. *Mon. Wea. Rev., 127,* 2204–2210. doi: 10.1175/1520-0493(1999)

Slatkin, B. (2015). *Effective Python: 59 specific ways to write better Python.* Upper Saddle River, NJ: Addison-Wesley.

Thompson, J. F., Soni, B. K., & Weatherill, N. P. (Eds.). (1999). *Handbook of grid generation.* Boca Raton, Fla: CRC Press.

9

MEANINGFUL VISUALS THROUGH DATA COMBINATION

Combining satellite imagery from multiple central wavelengths can help scientists detect clouds, vegetation, and air quality. Tutorials in this chapter show some simple examples of combined imagery, such as by using red bands (0.64 μm), near-infrared (0.86 μm), and window bands (10.3 μm), where atmospheric constituents minimally absorb radiation. Some examples include the construction of Normalized Difference Vegetation Index (NDVI), True Color imagery, and Dust RGB. The bands may not all have the same resolution or gridding, so this chapter will demonstrate some ways of handling data of different sizes.

Modern Earth observing satellites are equipped with multiple sensors that are sensitive to a range of wavelengths (also called **bands**) for visible, near infrared and short-wave infrared radiation. In remote sensing, these bands are also called **channels**. For simplicity, channels are sometimes referred to by either their **central wavelength** or an alias name based on a primary characteristic or application. For example, a 640 nm channel (the central wavelength) can have the alias "red channel" because it is sensitive over wavelengths in the red portion of the visible spectrum. However, the instrument is sensitive to radiation between 600–680 nm (Schmidt et al., 2017).

Some individual channels can be used to examine specific phenomena. For instance, IR channels can be used to study clouds, aerosols, snow cover, and atmospheric motion. Shortwave IR bands (e.g., 3.9 μm and 10.8 μm) can be used to identify fire hotspots, especially at night when there is less solar reflection to interfere with the signal. I can explore even more phenomena by combining or differencing channels. For instance, ice clouds can be detected by subtracting an 8.7

Earth Observation Using Python: A Practical Programming Guide, Special Publications 75, First Edition. Rebekah B. Esmaili.

DOI: 10.1002/9781119606925.ch9

μm channel from a 10.8 μm channel; water clouds are easier to distinguish by subtracting 12.0 μm from 10.8 μm. While single visible and IR channels are useful for identifying clouds, distinguishing the cloud types is difficult or impossible without data combination.

In addition to combining datasets, it is possible choose red-green-blue (RGB) color combinations to simulate human color perception. This is done by combining three satellite and assign one channel each to red, green, and blue tones. For a **natural color** (or **true color**) image, the channels will correspond to the actual red, green, and blue visible wavelengths. However, any channel can be assigned to red, green, or blue to create a create a **false color** (or **pseudo color**) image and make them perceivable to the human eye.

Examples in this chapter will build on what I learned using **NumPy, Pandas, NetCDF4, matplotlib.pyplot,** and **Cartopy**. In this chapter, I show examples using:

- **skimage**
- **sklean**

This chapter will heavily use Level 1b data (Section 3.3.1) from GOES-16 and -17, which are geostationary datasets. For brevity, I do not list them all, but these files have the following file structure:

- OR_ABI-L1b-RadM1-*

Additionally, the examples will use the cloud mask to filter clouds from our combined images as well as the downward shortwave radiation data to show an example of combining satellite data with ground observations:

- OR_ABI-L2-ACMM1-M6_G16_s20192091147504_e20192091147562_c20192091148155.nc
- OR_ABI-L2-DSRM1-M6_G16_s20192091300534_e20192091300591_c20192091303116.nc
- psu19209.dat
- surfrad_header.txt

9.1. Spectral and Spatial Characteristics of Different Sensors

While two different instruments may both have a "red" channel, instruments do not have a consistent spectral range or resolution because each mission and country have different needs and priorities. For example, Figure 9.1 shows some differences in channels for several instruments from several American (MODIS, VIIRS) and European (SLSTR, OLCI, MERIS) satellites. The red VIIRS channel has a central wavelength of 640 nm and a broader width (565–715 nm) than the MODIS red channel, which centers on 645 nm (620–670 nm). It is a good practice to check the bands when comparing results from different sensors. In addition to names, the channels are sometimes referenced by numbers, which are also inconsistent across satellite platforms. For example, the ABI channel 5 (a snow/ice band, 1.6 μm) will not necessarily be the same as channel 5 on Meteosat's SEVIRI

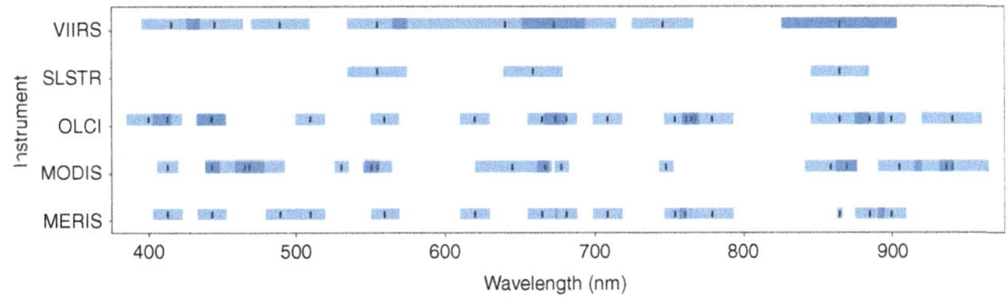

Figure 9.1 Example of various instrument channels and ranges between 400–1000 nm, which roughly encompasses visible and some near-IR wavelengths. Each vertical line is the central wavelength for the instrument which the shaded regions show where the bands are sensitive. Some channels have overlapping wavelengths, and these regions are darker.

(a water vapor band, 6.2 μm). To be as clear as possible, I will reference both the central wavelength and channel number in the text.

In addition to spectral differences, not all bands have the same spatial resolution. Visible imagery is typically higher resolution than IR channels because IR sensors require significant cooling to reduce noise and are thus more expensive to operate. Additionally, high resolutions mean larger data download from the satellite to the ground. As a result, we will apply simple gridding techniques from the previous chapter. This chapter will examine some ways to synergize measurements across sensors to visually detect Earth phenomena. While the examples in this chapter focus on geostationary satellites, the combination techniques can be used with imaging instruments on LEO satellites and CubeSats.

9.2. Normalized Difference Vegetation Index (NDVI)

Normalized Difference Vegetation Index (NDVI) is a means of distinguishing healthy vegetation and other surface phenomena. Chlorophyl absorbs highly in the red, while reflecting strongly in the near-IR. The difference is normalized by

$$NDVI = \frac{NIR - RED}{NIR + RED}$$

NDVI can range from –1 to 1. Healthy vegetation will have positive values, such as those found in forests and rainforests (values near 1) or shrubs and grassland (values from 0.2 to 0.4). Barren soil, rock, sand, and snow will have an NDVI close to zero because they reflect strongly in red. Negative values of NDVI indicate water. Clouds obscure the surface, so NDVI can be calculated using a cloud-free running mean of the surface over a period of days or weeks. Please note that NDVI is available as an official satellite product from NOAA (Vermote, 2018) and the following example is meant to be illustrative. The method below should not replace the official product, which has undergone additional peer review and quality control.

Below I import some of the now-familiar libraries: netCDF4 to open the relevant files, matplotlib to make plots, and NumPy for array manipulations.

```
from netCDF4 import Dataset
import matplotlib.pyplot as plt
import numpy as np, numpy.ma as ma
```

The next two lines of code import the radiances from Channel 2 (640 nm, `red`) and Channel 3 (860 nm, `veggie`) channels over the eastern United States on 28 July 28 2019, on a mostly clear day.

```
# Import ABI Channel 3
fname = 'data/goes-meso/ndvi/OR_ABI-L1b-RadM1-
M6C03_G16_s20192091147504_e20192091147562_
c20192091148025.nc'
goesnc = Dataset(fname, 'r')
veggie = goesnc.variables['Rad'][:]

# Import ABI Channel 2
fname = 'data/goes-meso/ndvi/OR_ABI-L1b-RadM1-
M6C02_G16_s20192091147504_e20192091147562_
c20192091147599.nc'
goesnc = Dataset(fname, 'r')
red = goesnc.variables['Rad'][:]
```

Comparing the dimensions of these two variables (using `red.shape` and `veggie.shape`), Channel 2 (0.5 km) has twice the resolution of the Channel 3 (1 km). In fact, Channel 2 has the finest spatial resolution of all ABI channels, because it is used to identify small-scale atmospheric features, like fog and overshooting cloud tops. Since I want to subtract the two channels, their array dimensions need to match. For simplicity, I will resize Channel 2 to decrease the dataset resolution by keeping only every other grid point. You can also implement another gridding scheme, such as those discussed in Section 8.3.

```
red = red[::2, ::2]
```

Now I can calculate the NDVI and create a plot using `imshow`. For simplicity, I do not include any cartographic features, which are discussed in greater detail in Chapter 7.

```
img = (veggie-red)/(veggie+red)

plt.figure(figsize=[12,12])
plt.imshow(img, vmin=-0.5, vmax=0.5, cmap=plt.get_cmap
("BrBG"))
plt.colorbar()
plt.show()
```

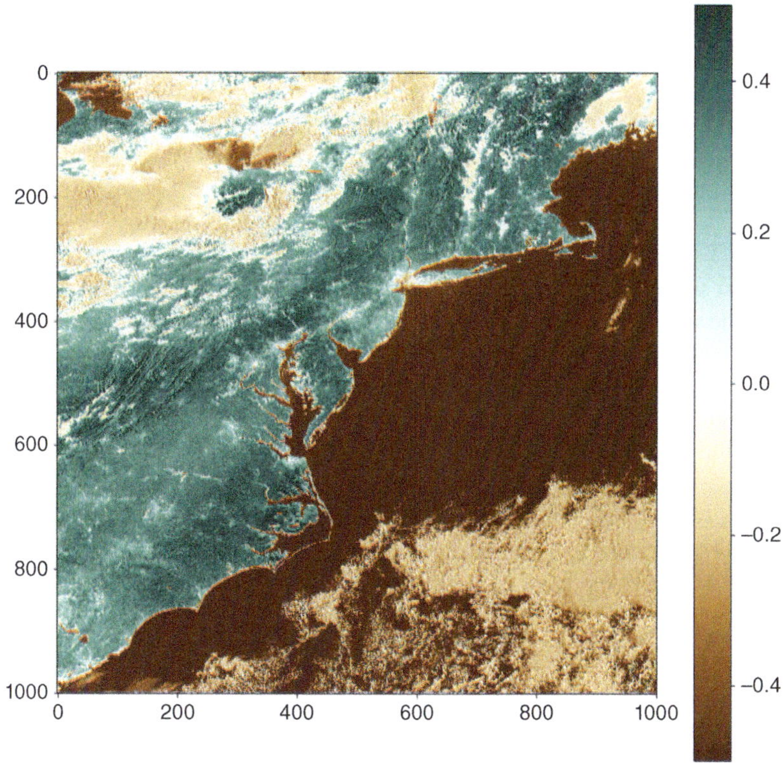

Plants are highly reflective and absorb strongly in the red bands, so high NDVI (green colored in plot) values indicate flourishing vegetation. This is particularly noticeable along the Appalachian mountain range. Large urban areas have very little vegetation and primarily reflect in the red bands but not much in the near IR. Water has low reflectance in the near IR, so over large bodies of water, the resulting index is very negative (dark brown).

While the plot above shows differences in regional vegetation, there is also visible cloud contamination with index values –0.2 and 0.0 (light brown). Clouds contain a significant amount of water, so they are not reflective in the near IR, and are highly reflective in the red band, so the final index value is slightly negative. However, this complicates imagery analysis because it the difference between clouds and barren soil is difficult to distinguish.

In order to remove the clouds, you can import the GOES AllSky product. The binary cloud mask (BCM) is 1 when a cloud is detected and 0 everywhere else.

```
fname = 'data/goes-meso/ndvi/OR_ABI-L2-ACMM1-
M6_G16_s20192091147504_e20192091147562_c20192091148155.nc'
goesnc = Dataset(fname, 'r')
cloud_mask = goesnc.variables['BCM'][:]
```

In contrast to the red channel, the binary mask is 4 km, and thus lower resolution than the veggie band. So, this time, I will have to upscale the data. To double the resolution, I will use the `np.repeat` command twice, once across `axis=1` to repeat values across the columns, and again across `axis=0` to repeat across rows. Again, more complex schemes might be necessary for analysis, but since I am visualizing the data, this will suffice.

```
cloud_mask_big = np.repeat(np.repeat(cloud_mask, 2,
axis=1), 2, axis=0)
```

To verify that the method worked, let's plot the cloud mask. Using a binary color scheme, black shows where clouds are detected (where `cloud_mask_big=1`) and white shows the clear sky (where `cloud_mask_big =0`).

```
plt.figure(figsize=[12,12])
plt.imshow(cloud_mask_big, cmap=plt.get_cmap("binary_r"))
plt.colorbar()
plt.show()
```

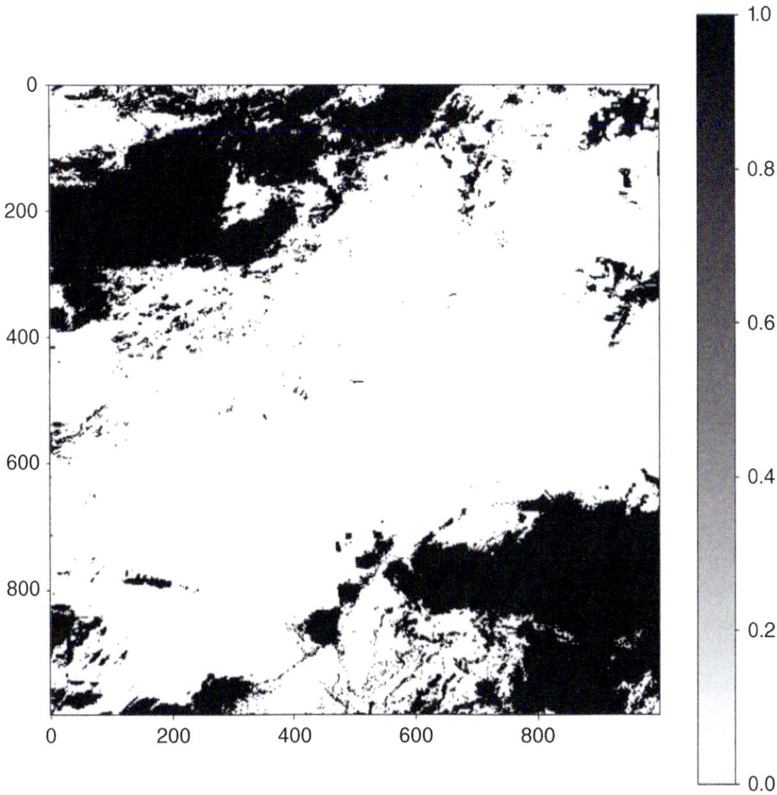

Masked arrays are useful for filtering unwanted features such as clouds (Section 5.2.1). Next, I convert the original NDVI image into a masked array using the binary cloud mask and set the fill value to `np.nan`.

```
imgMasked = ma.masked_array(img, mask=cloud_mask_big,
    fill_value=np.nan)
```

Finally, let's view the NDVI without cloud contamination. Since the clouds are masked out and the colorbar also has white values, I can update the background so that I can distinguish the clouds from where the NDVI is 0. To do this, use `plt.gca()` to update the current white background color to black.

```
plt.figure(figsize=[12,12])
plt.imshow(ma.filled(imgMasked), vmin=-0.5, vmax=0.5,
cmap=plt.get_cmap("BrBG"))
plt.gca().set_facecolor("black")
```

```
plt.colorbar()
plt.show()
```

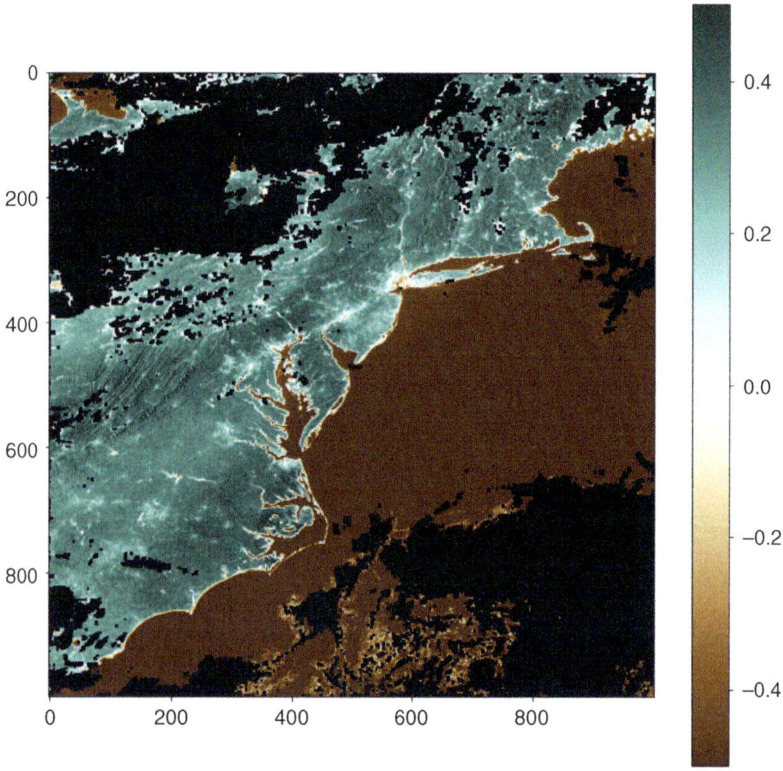

9.3. Window Channels

This "sandwich" imagery combines 0.64 μm (Red Band) visible and 10.3 μm ("Clean" IR Longwave Window Band) infrared imagery to distinguish cloud texture and show where cumulus towers are forming. Sandwich imagery is commonly used in operational weather forecasting to monitor for hazards.

The product is made by subtracting the IR band from the red channel:

$$Sandwich = Red - CleanIR$$

The brighter colors represent colder cloud tops, which have high red channel values because they are optically thick. Furthermore, clouds that are higher in the atmosphere have decreasing brightness temperature values. Thus, larger sandwich values indicate greater storm intensity.

```
# Import veggie channel (ABI Channel 13, 10.3)
fname = "data/goes-meso/michael/OR_ABI-L1b-RadM1-
M3C13_G16_s20182822019282_e20182822019350_
c20182822019384.nc"
g16nc = Dataset(fname, 'r')
C13 = g16nc.variables['Rad'][:]

# Import red channel (ABI Channel 2, .64)
fname = "data/goes-meso/michael/OR_ABI-L1b-RadM1-
M3C02_G16_s20182822019282_e20182822019339_
c20182822019374.nc"
g16nc = Dataset(fname, 'r')
C02 = g16nc.variables['Rad'][:]
C02 = C02[::4, ::4]

sandwich = C02 - C13
```

The following is mesoscale imagery of Hurricane Michael from 9 October 2019. The Clean IR channel is a quarter of the resolution of the red band (2 km versus 0.5 km), so I have to again resize Channel 2 to match the resolution of Channel 13.

Next, I can create the plot. I have selected the perceptually uniform `terrain` color map to help see the cloud texture.

```
plt.figure(figsize=[12,12])
plt.imshow(sandwich, cmap=plt.get_cmap("terrain"))
plt.colorbar()
plt.show()
```

9.4. RGB

RGB (red-blue-green) color combinations are useful because they create many colors perceptible to the human eye. In terms of satellite imagery, I am not limited to using red, green, and blue bands. I can also "see" other wavelengths of light by assigning their radiances to visible color scales. For example, I can create a plot where green is from 11.2 μm, a longwave IR wavelength. I can similarly construct a full-color image that acts to enhance or remove surface or atmospheric features based on the wavelengths they absorb or reflect in.

For RGB imagery, I will make use of gamma to rescale the radiance and brightness temperature values to 0 to 1 in order to convert them to colors.

```
from skimage.exposure import adjust_gamma,
rescale_intensity
```

Image enhancement also helps improve the utility of RGB imagery. Such enhancements include square root, equalized histogram, inverse hyperbolic sine functions, and gamma adjustments (Beh et al., 2018). Gamma rescales the luminance of an image following a power law. Using gamma values less than 1, it can brighten darker regions of an image; gamma values greater than 1 make dark colors become darker. Gamma can adjust the image so that changes are detectable by the human eye.

The next three sections examine the surface and atmospheric characteristics over the Alberta province in Canada on 30 May 2019. Alberta suffered from an extreme fire season that year, with over 2 million acres (880,000 hectares) burned. These fires led to large amounts of smoke that severely impacted the air quality in distant cities.

9.4.1. True Color

True Color images assign the actual red, blue, and green channels to the RGB constructed image. The ABI instrument from GOES-16 and GOES-17 have 0.47, 0.64, 0.86 µm channels, which also called the blue, red, and a near-IR "veggie" channel. The ABI does not have a green channel (~0.55 µm), but the green channel is linearly related to the values of blue, red, and veggie channels. Note that the green channel was included in on both SEVIRI and AHI respectively on the MSG and Himawari satellites.

Below, I import the blue and veggie channels. In addition, I use `adjust_gamma` to adjust all the channels by setting gamma to 0.5 to make it easier to perceive changes in values. Note that these adjustments are often determined through visual inspection and do not necessarily have any specific scientific origin.

```
# Import blue channel (ABI Channel 1)
fname = 'data/goes-meso/fires/OR_ABI-L1b-RadM2-
M6C01_G17_s20191501801013_e20191501801070_
c20191501801105.nc'
g17nc = Dataset(fname, 'r')
refl = g17nc.variables['Rad'][:]
blue = adjust_gamma(refl, 0.5)

# Import veggie channel (ABI Channel 3)
fname = 'data/goes-meso/fires/OR_ABI-L1b-RadM2-
M6C03_G17_s20191501801013_e20191501801070_
c20191501801103.nc'
g17nc = Dataset(fname, 'r')
refl = g17nc.variables['Rad'][:]
veggie = adjust_gamma(refl, 0.5)
```

As in the previous examples, the red channel's resolution is twice that of the blue and veggie channels. So below, I will resize the array to take every other pixel:

```
# Import red channel (ABI Channel 2)
fname = 'data/goes-meso/fires/OR_ABI-L1b-RadM2-
M6C02_G17_s20191501801013_e20191501801070_
c20191501801097.nc'
g17nc = Dataset(fname, 'r')
refl = g17nc.variables['Rad'][:]
refl = refl[::2, ::2]
red = adjust_gamma(refl, 0.5)
```

Now that the arrays are the same size, I can estimate the green radiance. Roughly speaking, the green "channel" is a linear combination of 45% of the Channel 2, and 45% of the Channel 1, and 1% of the Channel 3 (Lindstrom et al., 2017):

```
green = 0.45*red + 0.45*blue + 0.1*veggie
```

I can now combine the channels to produce the true color RGB. Since the arrays are imported as `masked_arrays` (Section 4.5), I can fill in the missing values with 0 (using `fill_value`) and then use the `np.stack` command to combine the images.

```
r = ma.filled(red, fill_value=0)
g = ma.filled(green, fill_value=0)
b = ma.filled(blue, fill_value=0)

rgb = np.stack([r, g, b], axis=2)
```

First, let me show the three images separately. Below, I am plotting each channel in a separate subplot. To make the data more comparable, I am setting the same `vmin` and `vmax` on each and using a grayscale colormap.

```
plt.subplot(131)
plt.imshow(r, vmin=5, vmax=25.5, cmap='tab20c')
plt.title("Red")
```

```
plt.subplot(132)
plt.imshow(g, vmin=5, vmax=25.5, cmap='tab20c')
plt.title("Green")

plt.subplot(133)
plt.imshow(b, vmin=5, vmax=25.5, cmap='tab20c')
plt.title("Blue")

plt.show()
```

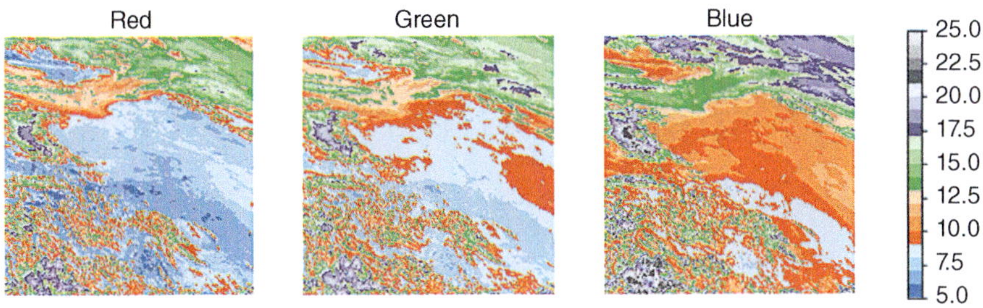

In the plots above, I used a high-contrast color scale so that the absorption differences across each channel are clearer. The magnitudes shown are rescaled radiance values, so only interpret them relative to each other. Larger values can be considered "brighter" or more "reflective" and while smaller values are "darker" or "more absorptive." Vegetation absorbs red wavelengths efficiently (shown as smaller values in the left-most plot) and is reflective in blue (larger values in the right-most plot). In general, clouds are highly reflective in all three wavelengths. However, water clouds will appear brighter in red and thick clouds will be brighter in blue (Berndt, 2017). By combining all three, I can better distinguish features on both the surface and in the atmosphere.

Below, I use the `rescale_intensity` to convert the radiance values, which range from 0 to 25.5 in this example, to a normalized, 0 to 1 scale. This is then plotted using `imshow`.

```
# Normalize values to 1
plt.figure(figsize=[12,12])
rgb255 = rescale_intensity(rgb, in_range=(0, 25.5),
out_range=(0, 1.0))
plt.imshow(rgb255)
plt.show()
```

Compared to the individual channels in the previous plot, it is easier to distinguish the land surface (darker colors) from the clouds (brighter colors) in the True Color image. You may notice that some clouds are bright while others are dull. The brightest clouds are thicker and contain ice and are thus more reflecting. These clouds will thus have high values in red, green, and blue to produce a brighter white. There are some dull gray clouds in the image as well, which result from the presence of smoke.

9.4.2. Dust RGB

Aside from very large dust events, which can regularly occur over the Saharan desert in Africa, dust plumes are more typically very thin and difficult to see in both visible and IR channels. However, channel differencing and gamma offset can be combined to improve visualization of some of the subtle differences in dust's absorption from other phenomena. In addition to detecting dust, Dust RGB can also be used to detect thick clouds and warm surfaces.

Dust RGB differences combinations of channels 11 (8.4 μm), 13 (10.3 μm), and 15 (12.3 μm), all of which are infrared. An advantage of using IR channels is that features are visible during the day and night. The formula is as follows:

$$Red = 12.3 - 10.3\ \mu m$$

$$Green = 10.3 - 8.4\ \mu m$$

$$Blue = 10.3\ \mu m$$

Using this formula, high values of red colors will show thick clouds or dust, green will identify water clouds, and blue will show warmer surfaces (Berndt, 2018). Large red values occur because dust absorbs more strongly at 10.3 μm than 12 μm, whereas the opposite is true for cold clouds. Lower clouds are warmer and will appear more blue, so dust will be identified through a combination of red and blue, giving dust a magenta color (EUMeTrain).

```
fname = 'data/goes-meso/fires/OR_ABI-L1b-RadM2-
M6C11_G17_s20191501801013_e20191501801070_
c20191501801110.nc'
g17nc = Dataset(fname, 'r')
btC11 = g17nc.variables['Rad'][:]

fname = 'data/goes-meso/fires/OR_ABI-L1b-RadM2-
M6C13_G17_s20191501801013_e20191501801070_
c20191501801113.nc'
g17nc = Dataset(fname, 'r')
btC13 = g17nc.variables['Rad'][:]

fname = 'data/goes-meso/fires/OR_ABI-L1b-RadM2-
M6C15_G17_s20191501801013_e20191501801076_
c20191501801112.nc'
g17nc = Dataset(fname, 'r')
btC15 = g17nc.variables['Rad'][:]
```

To provide some context to the code block below, channel 13 (10.3 μm, `blue`) is the "cleanest" of the three and least influenced by water vapor, so the other channels will be differenced from Channel 13. `red` is created by subtracting the dirty longwave IR (Channel 12) from the clean (Channel 13), so the color red implies the presence of moisture. `green` is the difference the 8.4 μm band (channel 11) and Channel 13, which isolates both moisture and sulfur dioxide (which is a byproduct of smoke).

```
img = btC15-btC13
# Rescale and adjust gamma
img = rescale_intensity(img, out_range=(0, 1))
red = adjust_gamma(img, 1.0)

img = btC13-btC11
# Rescale and adjust gamma
img = rescale_intensity(img, out_range=(0, 1))
green = adjust_gamma(img, 2.5)

img = btC13
# Rescale and adjust gamma
img = rescale_intensity(img, out_range=(0, 1))
blue = adjust_gamma(img, 1.0)
```

Like in the previous example, the intensity and gamma adjustment are made to each channel individually. The gamma adjustment will make all three images brighter, with the largest adjustment to green. As mentioned earlier, these adjustments enhance the channel differences, so that features are easier to see.

Below I fill the missing values and combine the RGB values using `np.stack`:

```
r = ma.filled(red, fill_value=0)
g = ma.filled(green, fill_value=0)
b = ma.filled(blue, fill_value=0)

rgb = np.stack([r, g, b], axis=2)
```

Finally, I can create the plot:

```
plt.figure(figsize=[12,12])
plt.imshow(rgb)
plt.show()
```

In the image above, cyan – a combination of blue and green – implies the atmosphere is clear but also marks the presence of sulfur dioxide, a component of smoke. So the presence of cyan above indicates that there is smoke. Clouds are the red/magenta colors and they obscure the underlying smoke. However, the smoke plume is visible between the clouds.

9.4.3. Fire/Natural RGB

Fire RGB, which is helpful for examining burn scars, combines the 1.6 μm (ABI Channel 5), 0.86 μm (ABI Channel 3), and 0.64 μm (ABI Channel 2) bands. Normally, healthy vegetation will be very green in this imagery. However, when vegetation dries out, the same area becomes a dark brown. This can occur during droughts after fires. Cyan colors show where there are ice clouds, while white indicates a liquid water cloud (EUMeTrain). The recipe, which was developed by EUMETSAT, is simply:

$$Red = 1.6\ \mu m$$

$$Green = 0.86\ \mu m$$

$$Blue = 0.64\ \mu m$$

Below I import the respective channels and rescale the brightness temperatures to a 0 to 1 scale:

```
fname = 'data/goes-meso/fires/OR_ABI-L1b-RadM2-
M6C05_G17_s20191501801013_e20191501801070_
c20191501801105.nc'
g17nc = Dataset(fname, 'r')
img = g17nc.variables['Rad'][:]
red = rescale_intensity(img, out_range=(0, 1))

fname = 'data/goes-meso/fires/OR_ABI-L1b-RadM2-
M6C03_G17_s20191501801013_e20191501801070_
c20191501801103.nc'
g17nc = Dataset(fname, 'r')
img = g17nc.variables['Rad'][:]
green = rescale_intensity(img, out_range=(0, 1))

fname = 'data/goes-meso/fires/OR_ABI-L1b-RadM2-
M6C02_G17_s20191501801013_e20191501801070_
c20191501801097.nc'
g17nc = Dataset(fname, 'r')
img = g17nc.variables['Rad'][:]
img = img[::2, ::2]
blue = rescale_intensity(img, out_range=(0, 1))
```

The above recipe does not require any gamma adjustments, so I can fill the masked array and combine the observations:

```
r = ma.filled(red, fill_value=0)
g = ma.filled(green, fill_value=0)
b = ma.filled(blue, fill_value=0)

rgb = np.stack([r, g, b], axis=2)
plt.figure(figsize=[12,12])
```

```
plt.imshow(rgb)
plt.show()
```

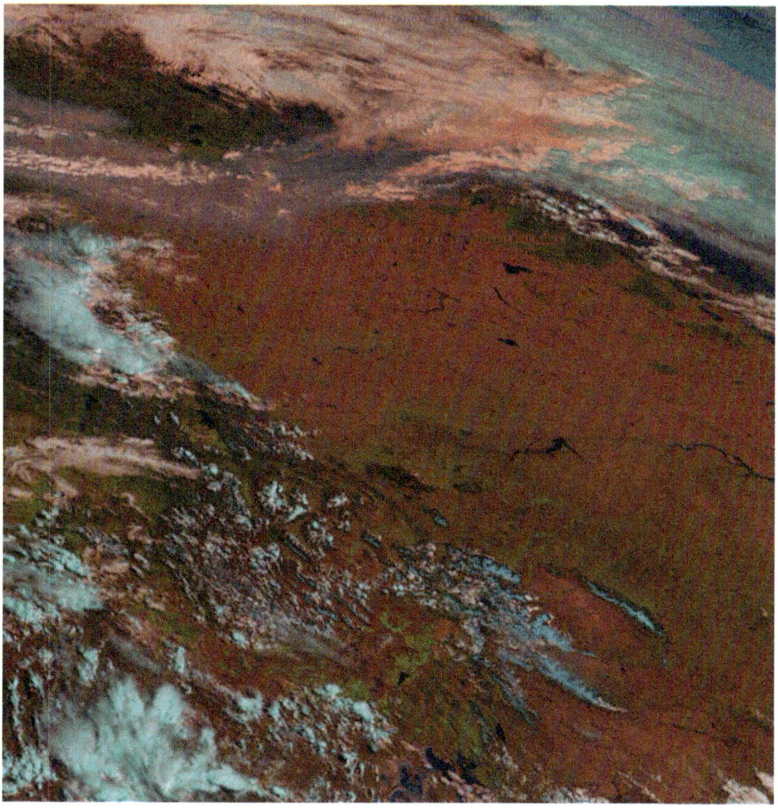

The image above shows large regions of brown, "burn scars" from recent fires. Ice clouds can be seen in the cyan tones. If you look at the Dust RGB image, this same region corresponds to thick clouds. This simplistic example illustrates how channel combination be used to identify surface and atmospheric features in a satellite imagery.

9.5. Matching with Surface Observations

Validation is the process of assessing satellite observations with other quality-controlled data, such as in situ observations (CEOS, 2016). Even for high-resolution satellites, a single footprint covers a large area of Earth's surface when compared to the "point observations" taken from a ground measurement. In this example, I will match GOES-16 Downward Solar Radiation (DSR) data with observations from a ground-based SURFRAD station at Pennsylvania State University (Penn State). The resulting collocated satellite and in-situ measurements are often called **match-ups**.

Below, I import the Cartopy library (Section 7.3.1) to create the plot from the GOES-16 DSR data. First, I create a `meshgrid` (Section 8.2) of the latitude and longitude coordinates.

```
import cartopy.crs as ccrs

fname = 'data/goes-meso/matchup/OR_ABI-L2-DSRM1-
M6_G16_s20192091300534_e20192091300591_c20192091303116.
nc'
goes = Dataset(fname, 'r')
dsr = goes.variables['DSR'][:]
lat = goes.variables['lat'][:]
lon = goes.variables['lon'][:]

x,y = np.meshgrid(lon, lat)
```

I will use `contourf` to plot the satellite data, and then create an `X` marker where the Penn State SURFRAD station is located (which per their website, is 40.72°N, 77.93°W)

```
plt.figure(figsize=[8,8])

ax=plt.axes(projection=ccrs.PlateCarree())
ax.coastlines('50m')

plt.contourf(x, y, dsr)
plt.colorbar()

plt.scatter(-77.93, 40.72, marker='x', c='black', s=100)
ax.set_ylim(34, 45)
ax.set_xlim(-80, -70)

plt.show()
```

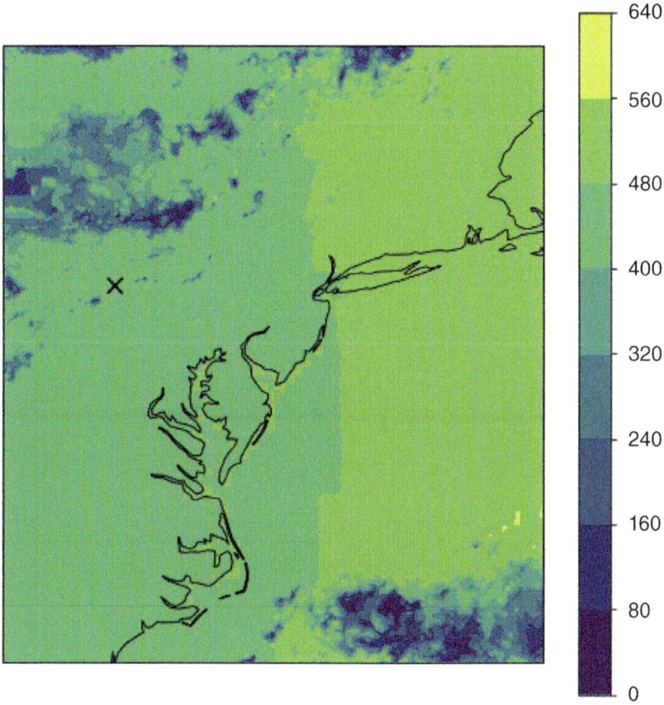

In the next two sections, I show two methods for performing matchups: The first only uses NumPy and the haversine formula, while the second uses `DistanceMetric` from sklean.neighbors, which utilizes a machine-learning based search. The first method is helpful if you want to limit the number of package dependencies in your code. The second method gives you an opportunity to simplify your code and work with a powerful data science tool.

9.5.1. With User-Defined Functions

This method is useful because it relies solely on NumPy functions. I create two functions (Section 4.9) that will help us perform the match-ups. Writing this code as a function is optional, but it gives you an opportunity to practice. The haversine formula calculates the distance between two points on the sphere. In the `haversine` function below, the inputs are the latitude and longitude of the retrieval point (`deglat1` and `deglon1`) and the station point (`deglat2` and `deglon2`), and it return the distance (`dist`) in kilometers. When writing function, I recommend adding comments about the purpose of the code (Section 12.2).

```
def haversine(deglat1, deglon1, deglat2, deglon2):
'''This function uses the haversine formula to calculate the
distance between two latitude and longitude coordinates and
returns the distance in km.'''

    r_earth = 6378.0

    lat1 = deglat1*np.pi/180.0
    lat2 = deglat2*np.pi/180.0

    long1 = deglon1*np.pi/180.0
    long2 = deglon2*np.pi/180.0

    a = np.sin(0.5 * (lat2 - lat1))
    b = np.sin(0.5 * (long2 - long1))

   # Result in radians, multiply times the diameter of the
earth (6,356 km)
   dist = r_earth * 2.0 * np.arcsin(np.sqrt(a * a + np.cos
(lat1) * np.cos(lat2) * b * b))

    return dist
```

Rather than search every point for a match, you can limit the number of computations by only searching a smaller subset of points, such as all points closer than 1°. Below, I define a function (`matchup_spatial`) to take a list of latitude and longitude pairs and find all matches within a given radius. If the radius is not supplied, a default value of 50 km will be passed in. This function will return a `mask` that is `True` when values are greater than the radius and `False` when they are within the radius.

```
def matchup_spatial(latitude, longitude, site_lat,
        site_lon, radius_km=50.0, closest_only=False):
'''This function calculates the distance between a list of
retrieval coordinates and a point observation. It returns all
matches that are within a given radius or the closest
point.'''

    # Find index for pixels within radius_km around ground
site
    distance_matrix = np.full(latitude.shape, 6378.0)
```

```
#Calculate the distance in degrees
dist_deg = np.sqrt((np.array(latitude)-site_lat)**2
     + (np.array(longitude)-site_lon)**2)
close_pts = (dist_deg < 1.0)

# Replace angle distance with km distance
distance_matrix[close_pts] = haversine(site_lat,
      site_lon, latitude[close_pts],
      longitude[close_pts])
keep_index = (distance_matrix > radius_km)

# Return a single (closest) value
# if closest_only:
#    if len(keep_index[keep_index==True]) > 0:
#          keep_index = (distance_matrix ==
#                distance_matrix.min())

return keep_index
```

The first line creates a matrix that is filled (`np.full`) with a large value (`6378`, for the radius of the Earth in km).

In the next step, I create an array that calculates the geolocation differences between the point and the array of retrieved values. This step is followed by a mask that keeps only the points that are less than 1° away. This will reduce the number of times that I call the `haversine`, which can speed up my code. At the equator, 1° is approximately 111 km. If you want to include search radiuses greater than this, you will need to increase the value in `index = (mask < 1.0)`.

The third code block calls `haversine` for only the closest values. If the distance is outside of the radius, an index is returned.

There are a few lines of code commented out above. In some cases, you will want all the points within the radius but at other times, you may want the closest point. For the latter case, uncommenting the last code block will only return the closest match.

Now I can call the functions that I created. I pass in `x` and `y`, the coordinates of the SURFRAD station, and increase the search radius to 100 km.

```
mask = matchup_spatial(y, x, 40.72, -77.93, radius_km = 100.0)
```

The `matchup_spatial` function returns a mask, which I will apply to the `dsr`, `x`, and `y` so that I will only plot the points within 100 km of the stations:

```
dsrMA = np.ma.masked_array(dsr, mask=mask, fill_value=0)
xMA = np.ma.masked_array(x, mask=mask, fill_value=0)
yMA = np.ma.masked_array(y, mask=mask, fill_value=0)
```

Now, I can plot the matches and see if the points I selected are visually within range of the station.

```
plt.figure(figsize=[8,8])
ax=plt.axes(projection=ccrs.PlateCarree())
ax.coastlines('50m')
plt.contourf(xMA, yMA, dsrMA)
plt.colorbar()
plt.scatter(-77.93, 40.72, marker='x', c='black', s=100)

ax.set_ylim(32, 45)
ax.set_xlim(-80, -70)

plt.show()
```

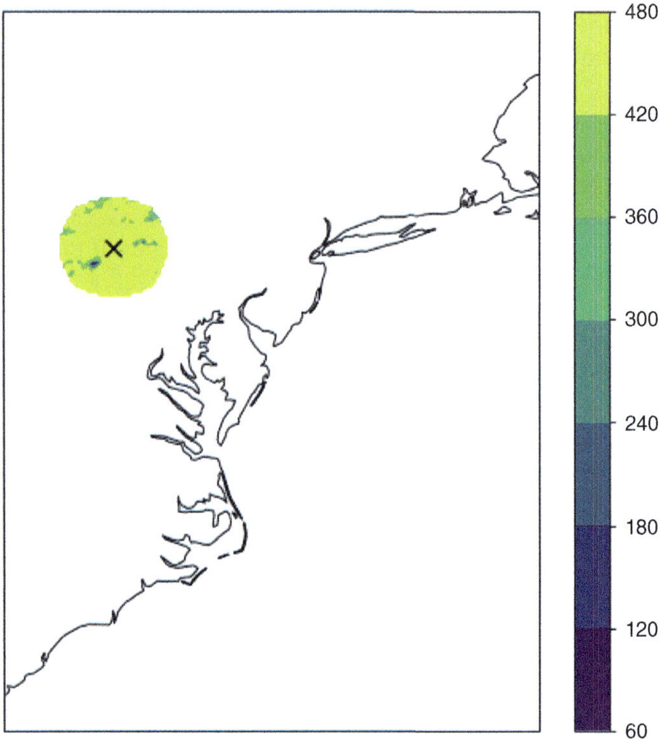

From the plot above, I can visually confirm that the data are within 100 km of the site.

Now that the satellite data are imported and I have found the closest spatial points, next I will import the surface data. SURFRAD collects observations every 10 minutes and there will only be one time match. SURFRAD observations are saved in text files, so I use Pandas here because it will easily import the observations into a DataFrame. There are two import statements because the data and column names are stored in separate files, which I will call the data and header files. I will use `read_csv` to import the header file (the first import statement), which I save to the `header` variable. I then parse the columns of the data file using the `header=None` and `names=list(header)` option in the second `read_csv` statement below. If you inspect data file using Excel or a text editor, you can see that I need to skip the first two rows (`skiprows=2`) and the data are separated by whitespace, not commas (`delim_whitespace=True`).

```
import pandas as pd

fname = 'data/goes-meso/matchup/surfrad_header.txt'
header = pd.read_csv(fname)

fname = 'data/goes-meso/matchup/psu19209.dat'
ground = pd.read_csv(fname, skipinitialspace=True,
     delim_whitespace=True,
     skiprows=2, header=None, names=list(header))
```

Now I also need to consider which SURFRAD points are closest in time to the DSR observations. First, I prepare the SURFRAD data by converting the time columns to a single column in the datetime format (Section 4.6) for easier handling.

```
# Surface obs time
df = pd.DataFrame({'year' : ground['year'],
      'month' : ground['month'], 'day' : ground['day'],
      'hour' : ground['hour'], 'minute' : ground['min']})

# List of ground observation times
ground['Datetime'] = pd.to_datetime(df )
```

Next, I extract the time from the DSR file. I will use this satellite time to search the SURFRAD time array.

```
# Satellite observation time
fmt = '%Y-%m-%dT%H:%M'
fileTime = pd.to_datetime(goes.time_coverage_start[0:19],
format=fmt)
```

Below I create another function (matchup_temporal), which is similar to the spatial match-up, but it will only return a single time match.

The inputs to the function are the time I wish to match with (time), an array of times (time_array), and the maximum time different in minutes (matchup_max_time_mins), with a default value of 15 minutes. This is a reasonable value for this example, but the exact match-up times differences will vary by application. In the first line, I take the difference between time and time_array. I use the list comprehension syntax (Section 4.8) to iterate over time_array. The index variable picks the minimum time difference value from the previous computed line.

The third line calculates the absolute values of the time difference. Using an if statement, index is returned if the time difference is less than 15 mins away. Otherwise, –1 is returned if no match is found.

```
def matchup_temporal(time, time_array,
    matchup_max_time_mins=15):

    time_diff = [np.abs(x - time) for x in time_array]
    index = np.argmin(time_diff )

   ifabs(time_array[index]-time)<pd.Timedelta(minutes=15):
        return index
   else:
        return -1
```

Now, I can perform the temporal search. I will print the index to make sure a match was found, which will be –1 if there is no match:

```
index = matchup_temporal(fileTime, ground['Datetime'] )
print(index)
781
```

Finally, I can compare the values. Below I print the average value observed around the site using the satellite.

```
ground['dw_solar'][index] - dsrMA.mean()
25.645798230834032
```

This yields a difference of 25.64 W/m2 for this example. In practice, this process will be repeated over several sites, days, and times or be part of continuous monitoring. This is why I recommend writing these methods into simple functions will make scripting this data easier, which I discuss more in Section 11.1.

9.5.2. With Machine Learning

Scikit Learn (Pedregosa et al., 2011) is a robust machine-learning package for Python. While exploring machine-learning capabilities would fill another book, there are some aspects that make it a useful tool for searching gridded satellite data. Scikit Learn's *neighbors* module uses the *k*-nearest neighbors algorithm to find the closest match between datasets, which includes physical distances between georeferenced points (using the *DistanceMetric* method). The haversine method that I used in the previous section is one of the available distance metrics.

I will repeat the example from the previous section finding all DSR retrievals within 100 km of the Penn State ground station. This time, I will use the scikit learn to compare the differences in approaches. To begin, I import the necessary method below:

```
from sklearn.neighbors import DistanceMetric
```

Later in this example, I will use the `pairwise()` method to compute the distance between the array of latitude and longitude coordinates. Scikit Learn's haversine function requires the points to be ordered as a list within of list of [latitude, longitude] pairs. Additionally, all inputs and outputs are expressed as radians. If you recall the haversine function that I defined in previous section, I also converted the coordinates to radians. To first prepare the data, construct a list of points (`samples`) from the DSR dataset (which were stored in the `lat` and `lon` – you need to import the data in the first code block in the previous section) using a list comprehension (Section 4.8). The pairwise method is also able to calculate the haversine distance between an array of points; however, since I am only comparing the distance to a single point, I will only nest this single point in another list (making it a list with length 1).

```
samples = [[i, j] for i in lat for j in lon]
point = [[40.72, -77.93]]
```

Next I will convert the individual items into radians using `np.radians()`. This procedure must be done twice, once for the `samples` variable and once for `point`. To simplify the code, I define a function below. Then, I run this function to convert `points` to radians (`points_rad`) and `samples` to radians (`samples_rad`).

```
# Convert to radians
def deg_to_rad(pairs):
    return [[np.radians(x[0]), np.radians(x[1])] for x in
pairs]

point_rad = deg_to_rad(point)
samples_rad = deg_to_rad(samples)
```

Now that I have prepared the data, I will use the `DistanceMetric.get_metric('haversine')` method to access the haversine function, which I define in the first line. The next line uses the `pairwise()` method to calculate the distance between `point_rad` and `samples_rad`. I multiply these values by the radius of the Earth (6,378 km) to convert from radians back to kilometers. The resulting array is a list of distances (`distance`) of each satellite retrieval latitude and longitude coordinates and the point. The `distance` array has a size of (1, 501720), so I only extract the first element subsetting along the 0 index (`distance[0]`). The third line of code (`distance_2d`) uses the `reshape` command to reformat the data so that it matches the dimensions (888, 565) of the two-dimensional latitude, longitude, and DSR variables (using NumPy's `.shape` attribute).

```
dist = DistanceMetric.get_metric('haversine')

distances= dist.pairwise(point_rad, samples_rad)[0]*6378.0
distances_2d = distances.reshape(dsr.shape)
```

The remainder of the example resembles the same process that I used in the previous section. I create a mask to show which values I do not want in the final dataset (which are those further than 100 km from the given point) and then create masked arrays to exclude these values from any further figures and analysis.

```
mask_sk2d = distances_2d > 100

dsrMA_sk = np.ma.masked_array(dsr, mask=mask_sk2d)
xMA_sk = np.ma.masked_array(x, mask=mask_sk2d)
yMA_sk = np.ma.masked_array(y, mask=mask_sk2d)
```

Below I plot a comparison with the distance calculation (`distanced_2d`) and the matched results using `plt.subplots` (Section 6.3.6). Since both plots will be placed on the same map, these elements are iterated over all axes to simplify the code.

```
proj = dict(projection=ccrs.PlateCarree())
fig, axes = plt.subplots(ncols = 2, figsize=[10,8],
subplot_kw=proj)

axes[0].set_title("Distance from "+str(point[0]))
axes[0].pcolormesh(x, y, distances_2d, vmin=0, vmax=900)

axes[1].set_title("Matches < 100 km from "+str(point[0]))
axes[1].contourf(xMA_sk, yMA_sk, dsrMA_sk)

for axis in axes:
    axis.coastlines('50m')
    axis.set_ylim(32, 45)
    axis.set_xlim(-80, -70)
    axis.scatter(-77.93, 40.72, marker='x', c='black',
          s=100)

plt.show()
```

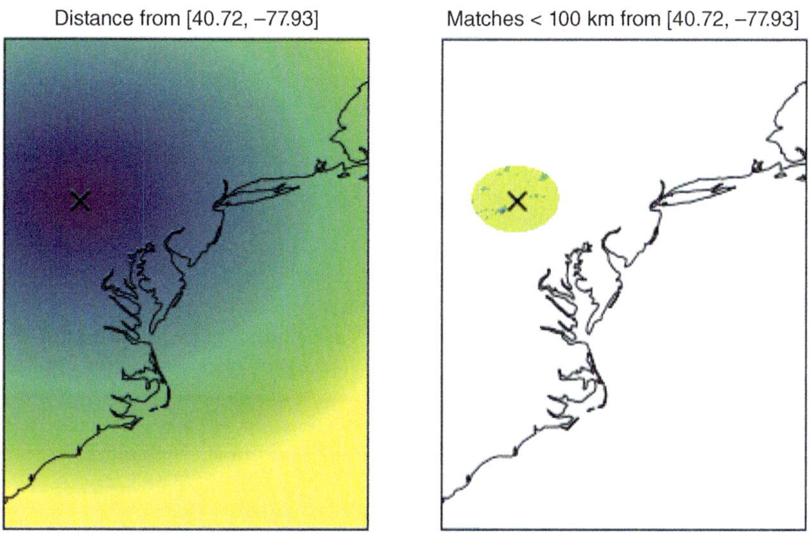

As expected, the lowest values are closer to the Penn State station (denoted by the X) and the matches on the right are within the 100 km criteria. The Scikit Learn method of performing match-ups has fewer lines of code when compared with the first method. However, if you want to limit the number of package dependencies in your code, you may want to define your own functions and techniques, as I did in the previous section.

9.6. Summary

Combining different imagery is useful for monitoring changes in vegetation using NDVI, while TrueColor, Dust RGB and Fire RGB can show cloud and surface features. There are significantly more RGB recipes than those presented in this chapter. For practice, I encourage you to explore more. The SatPy package (which was developed by the PyTroll community, who also maintain pyresample) has many of these composite recipes built in. If you are interested in learning this package, the group has some excellent examples online: https://satpy.readthedocs.io/en/latest/examples/index.html. Satellite datasets are also frequently compared with other satellite, model, and in-situ data for scientific research and hazard monitoring. To make these comparisons, you may have to transform the data's grid (such as using the techniques in Chapter 8) or using search algorithms to find the closest points, just as you saw in this chapter.

In the last few chapters, you have learned some basic techniques for working with satellite data and creating visualizations. In the next section, I will discuss how to export your valuable analysis and high-quality figures.

References

EUMeTrain. (n.d.). RGB colour interpretation guide. http://www.eumetrain.org/RGBguide/rgbs.html

Pedregosa, F., Varoquaux, G., Gramfort, A., Michel, V., Thirion, B., Grisel, O., et al. (2011). Scikit-learn: Machine learning in Python. *Journal of Machine Learning Research, 12*(85), 2825–2830.

Schmit, T. J., Griffith, P., Gunshor, M. M., Daniels, J. M., Goodman, S. J., & Lebair, W. J. (2017). A closer look at the ABI on the GOES-R series. *Bulletin of the American Meteorological Society, 98*(4), 681–698. https://doi.org/10.1175/BAMS-D-15-00230.1

Working Group on Calibration & Validation (WGCV): Terms of reference (ToR) version 1.0. (2016, November 2). CEOS. http://ceos.org/document_management/Publications/Governing_Docs/WGCV_ToR-v1.0_02Nov2016.pdf

Berndt, E. (2016, November 14). Natural color/FIRES RGB quick guide. NASA/SPoRT. https://weather.msfc.nasa.gov/sport/training/quickGuides/rgb/QuickGuide_NatColor-Fire_NASA_SPoRT.pdf

Berndt, E. (2017, September 19). Day land cloud rgb quick guide. NASA/SPoRT. https://weather.msfc.nasa.gov/sport/training/quickGuides/rgb/QuickGuide_NatColorFire_-NASA_SPoRT.pdf

Lindstrom, S., Kaba, B., Schmidt, T., & Kohrs, R. (2017, October 3). *CIMSS natural true color quick guide*. University of Wisconsin. Retrieved from http://cimss.ssec.wisc.edu/goes/OCLOFactSheetPDFs/ABIQuickGuide_CIMSSRGB_v2.pdf

Vermote, E., & NOAA CDR program. (2018). *NOAA climate data record (CDR) of AVHRR Normalized Difference Vegetation Index (NDVI)*, Version 5 [Data set]. NOAA National Centers for Environmental Information. https://doi.org/10.7289/V5ZG6QH9

10

EXPORTING WITH EASE

Of increasing importance, researchers are now required to archive their data after it has undergone analysis and visualization. Some journals and grants now require scientists to make their data publicly available to ensure that the conclusions are reproducible and transparent. This chapter will show some common ways to store data, whether it be as publication-quality images, text files, or netCDF4. For self-describing data, this chapter provides an overview of some important conventions to make data easier to distribute, share, and archive.

It is very likely that your Python projects will generate data or imagery that you will want to save for future use, either for yourself or to share with others. Furthermore, many public grants and research journals require authors to save data for five years. There are numerous formats you can use to save your data and visualizations using Python. Plots can be saved in jpeg, png, or pdf. Binary files (Section 3.2.1) may not be human readable, but they have good compression and are easy to import and manipulate using packages like **NumPy.** Text files (like csv or tsv, Section 3.2.2) are adequate when the dataset is small and needs to be readable by others. However, neither are good for long-term storage nor sharing. As discussed in Section 3.2.3, **netCDF** and **HDF** are useful for long-term archiving of results, are cross platform, and are readable outside of the Python environment. These formats support metadata storage and can be read on MacOs, Windows, and Linux operating systems. Self-describing formats are easier to disseminate within your research community, and if you add descripting metadata, you can more easily return to your results after extended periods of time.

Earth Observation Using Python: A Practical Programming Guide, Special Publications 75, First Edition. Rebekah B. Esmaili.

DOI: 10.1002/9781119606925.ch10

In this chapter, I will show how you can export data to text, pickle, and netCDF using common packages. The examples in this chapter will use **NumPy, Pandas, netCDF4,** and **xarray,** which you have learned about extensively in other chapters. Not only do these packages allow you to import data, but they can also save data as well. I will also introduce two new packages, **csv** and **pickle**, which can archive text and binary formatted, data, respectively. Figures can be saved using `matplotlib.pyplot`.

10.1. Figures

Saving plots is relatively straightforward. In Chapter 6, you learned to end matplotlib code segments with `plt.show()`. To write the image to file, I use the `plt.savefig()` command, followed by `plt.close()`.

```
plt.figure(figsize=[5,5])
plt.hist(fires['frp(MW)'],  bins=bins10MW)

plt.savefig('histogram.png',  bbox_inches='tight')
plt.close()
```

The `plt.close()` command tells Jupyter Notebook to run in **noninteractive mode**, so no figure will render. Instead, a png image named `histogram.png` is saved in your default directory, which is where you launched Jupyter Notebook by default. If you want to save the image somewhere else, you can include either an absolute or local relative file path before the filename. `plt.savefig` will automatically detect the file format from the extension. The second argument (`bbox_inches`) refers to the whitespace around the plot, so it is aesthetic and optional. If you leave it out, the canvas will have a lot of whitespace surrounding the plot. Adding `bbox_inches='tight'` option will fit the canvas to the plot more closely. You can alternatively specify the whitespace you want. For example, `bbox_inches=[1,1]` will add 1 inch of whitespace around the plot. Note that the tight layout only includes tick labels, axis labels, and titles. If you have other elements outside of the plot, they may overlap or be cropped from the image.

Image file formats like PNG, JPG, and GIF are **rasterized** – that is, they hold information in pixel space (Table 10.1). JPG and PNG formats are better for images with many colors. PNG is a good choice because it can support unlimited colors, with a moderate file size. If there are only a few colors (such as for line or scatter plots), GIF are a good choice because the file size is the smallest.

If you wish to create simple animations, GIF can easily be converted into animated GIFs either through online or software tools. For publication quality figures, **vector** images are preferred because they can be zoomed in indefinitely without a loss of quality and usually have smaller file sizes. The most common

Table 10.1 Differences between Image Types

Characteristic	JPG	GIF	PNG
colors	millions	256	unlimited
compression	lossy	lossless	lossless
transparency	no	yes	yes
animation	no	yes	no
file size	large	small	medium

formats are SVG and PDF. Today, many journals require authors to submit their figures as vector graphics. I recommend PDF because they are platform independent and there are many tools to read them, including most web browsers. PDFs can be larger than PNGs for complex satellite imagery. If storage space is not a limitation for you, it might be helpful to save the figure in both a raster and vectorized format.

10.2. Text Files

I briefly looked at the `open` command back in Section 5.1.1. I described it as a useful method of opening an irregular text file for reading. I can also use the same command to write files as well. `Open` can write both text and binary files, but below I will only show a text example since there are simpler methods for saving binary files (which I'll show in the next section). Suppose I have three lists of items to write to file:

```
name = ['GOES-16', 'IceSat-2', 'Himawari']
agency = ['NOAA', 'NASA', 'JAXA']
orbit = ['GEO', 'LEO', 'GEO']
```

The following code is a loop and must be run all at once. I first import the csv library to facilitate writing the file. The `writer` command defines the file and how to separate the data. Then `writerow` assigns the data. You can pass either a single string of text (as I did with the headers), or create a loop to write items from a list to the file.

```
import csv

with open('satellite.txt', 'w', newline='') as csvfile:
    writer = csv.writer(csvfile, delimiter=',')

    # adds a header
    writer.writerow(['name', 'agency', 'orbit'])
```

```
# Writes each element as a single row
for i in range(0, len(name)):
  writer.writerow([name[i], agency[i], orbit[i]])
```

The above command is useful if you want to have significant control over the layout and formatting of the final file. If you want to quickly save a readable file, using the `to_csv` command in Pandas is convenient. The option `index=False` suppresses the indices of the DataFrame (which are printed to the left of the DataFrame) from being printed to file.

```
import pandas as pd
df = pd.DataFrame({'name': name,'agency': agency,
'orbit': orbit})

df.to_csv('satellites.csv', index=False)
```

10.3. Pickling

In Python, **pickling** is a method of storing serialized binary data for later use. Pickle files (`.pkl`) can store any type of object. Pickling gets its name from the culinary practice of salting food to preserve and store for long periods of time.

Below, I import the **pickle** library and use the `pkl.dump` command by passing the DataFrame name (note that any object can be pickled) and create a new file called `satellites.p` using open. Instead of `'w'`, I use `'wb'` to indicate write binary.

```
import pickle as pkl
pkl.dump(satellites, open("satellites.p", "wb"))
```

The file should be saved in your default location. If I want to open the saved file, I use the open command with `'rb'` (read binary) and `pkl.load`:

```
satellites2 = pkl.load(open("satellites.p", "rb"))
```

10.4. NumPy Binary Files

Like pickle files, NumPy binary files (`.npz`) are also binary files but more geared toward arrays, which can be multidimensional. You can also conveniently can store multiple datasets inside a single file. For instance, I will save both the name and orbit in a file called `satnames.npz` (the extension is automatically added). The `name=`, `agency=`, and `orbit=` are arbitrary names and optional.

However, I recommend including them because otherwise NumPy will assign generic labels to the data:

```
import numpy as np
np.savez('satnames', name=name, agency=agency,
orbit=orbit)
```

Now, I can reopen the file using the `np.load` command and assign it to a variable called `npzfiles`. I can print the available arrays inside `npzfile` using the `.files` attribute. Since I stored two arrays inside the file, this returns two values, name and orbit. Then use the `npzfile['name']` command to access the list:

```
npzfile = np.load('satnames.npz')
npzfile.files
['name', 'agency', 'orbit']

npzfile['name']
array(['GOES-16', 'IceSat-2', 'Himawari'],
dtype='<U8')
```

For a typical scientific project, I likely already have NumPy imported. Thus, using the npz format over pkl means I have one fewer dependency. In terms of file size, they are not significantly different. However, they are best used as **intermediate files**, which are data saved prior to the completion of the main program. Intermediate files can help speed up processing if there are calculations that do not need to be dynamically calculated. However, I do not recommend either format for data that you wish to share because pkl and npz are not self-describing formats and are not easily read without Python.

10.5. NetCDF

Because they are so prevalent in Earth science, much of this book discusses how to read and understand the structure of netCDF files. Similar to reading netCDF files, the process of creating a new netCDF file can be tedious, but it is worth investing the time to learn how the files can be organized. To accomplish this, I will first use the more verbose netCDF4 package. However, afterward, I will demonstrate how you can quickly (and with fewer lines of code) create netCDF files using xarray.

10.5.1. Using netCDF4 to Create netCDF Files

In the following example, I will import multiple JPSS AOD files and combine them. Swath data are often stored into one-minute scans, and it can be helpful to combine them. Additionally, I will quality control the data using the quality flag variable.

So, first I import the requisite libraries. We have not used glob package in this text, glob will return a list of files that contain a given string or character pattern. You will notice that the files we are searching for end in asterisk (*). This tells to the command to search for netCDF files that begin with JRR-AOD, regardless of what character pattern that follows.

```
from netCDF4 import Dataset
import numpy as np
from glob import glob
```

Next, I search for all files in the directory:

```
# Get file names, exit if no files are found
filenames = glob('data/aod/JRR-AOD*.nc')

# Import datasets
file_id = Dataset(filenames[0], mode='r')
```

I want to create a new netCDF file, so I need to know the dimensions of the final dataset. There are many variables in the file, but since I want to save just the best quality AOD550, the only relevant variables are `Latitude`, `Longitude`, `AOD550`, and `QCAll`. I can ignore the rest to help reduce the file size. Below, I am printing dimensions names, and I will show that they are all the same:

```
for var in ["Latitude", "Longitude", "AOD550", "QCAll"]:
  print(var, file_id.variables[var].dimensions)
Latitude ('Rows', 'Columns')
Longitude ('Rows', 'Columns')
AOD550 ('Rows', 'Columns')
QCAll ('Rows', 'Columns')
```

Since I am combining three files, I will have a final dataset that is bigger than the original in one dimension. Below I make a list of the dimension sizes in the first file. In this example, all three files have the same dimension sizes. This is a characteristic of many VIIRS and ABI products, but will not be true of all datasets.

```
for dim in ['Columns', 'Rows']:
  print(dim, len(file_id.dimensions[dim]))
Columns 3200
Rows 768
```

Next, let's import the data above into a variable and count the number of files:

```
cols = len(file_id.dimensions['Columns'])
rows = len(file_id.dimensions['Rows'])
```

```
num_files = len(filenames)
file_id.close()

print(rows, cols, num_files)
768 3200 3
```

Now that I have information on the size, I can create the netCDF file. I will use the `Dataset` module from the netcdf4 package but will use the `"w"` argument to write rather than read (which, if you recall, uses the "r" argument). I write a general description of what is in the data using `.description`, which assigns the text string to the global metadata.

```
# Create output file name
output_filename = "JRR-AOD-combined.nc"

# Open a new netCDF file
rootgrp = Dataset(output_filename, "w", format="NETCDF4")
rootgrp.description = "Combined NOAA-20 AOD swaths for
Alberta Fire"
```

Before I can fill the dataset, I need to create the dimensions (`.createDimensions`) and variables (`.createVariables`). Because swath data are scanned along the north–south direction, I will concatenate the data in the row dimension. The size of the column dimension will remain unchanged but the latitude dimension will be three times the original size. Note that if the files had irregular dimensions, I would have to read in all the files to get the individual dimension sizes and potentially grid them to a new multidimensional array using a gridding technique from Chapter 8.

```
# Format the latitude and longitude variables
lat = rootgrp.createDimension("lat", rows*num_files)
lon = rootgrp.createDimension("lon", cols)
```

The two lines above define the dimensions, `lat` and `lon`. Note that some other datasets may have time and height as well. When creating the latitude and longitude dimensions, you must provide single integer value length, such are `rows*num_files` and `cols`, respectively. Next, I create the variables:

```
latitudes = rootgrp.createVariable("Latitude","f4",
("lat","lon"), \
zlib = True, least_significant_digit = 2)
longitudes = rootgrp.createVariable("Longitude","f4",
("lat","lon"), \
zlib = True, least_significant_digit = 2)
```

```
variable = rootgrp.createVariable('AOD550', "f4",
("lat","lon"), \
zlib = True, least_significant_digit = 2, fill_value =
-999.9)
```

These three lines are the placeholder for the latitude, longitude, and variable (which is AOD). When I create the variable, I will create a two-dimensional array since the coordinates are irregularly spaced. If the data were evenly spaced, then a single-valued array would suffice for latitude and longitude, although AOD would still be two-dimensional. I assign the variable a name, as it will appear inside the netCDF file, along with a number type, and the two coordinate dimensions. The last two keywords are optional but are for compressing the dataset.

If you are a Linux user, you may be familiar with gzip and tar, which are the primary compression and archiving utilities. Gzip and tar are both wrappers for zlib, a free data compression library that uses an algorithm to "deflate" the data to reduce the file size. For text, zlib can reduce file sizes up to 50%, although binary-based datasets have lower compression rates. Without zlib=True in the code above, the final netCDF file size will be much larger.

The option least_significant_digit is for zlib, which indicates the smallest decimal place of the variable that is precise. In the example, I use a value of 2 so that the data are saved to two decimal places. Smaller values indicate more compression (reducing file size) but will increase the time required to save the file. The opposite is true for larger values.

In the variable group for AOD550, I also assign a fill_value (Section 3.2.3) of -999.9 for indicating where the data should be ignored, because there was no retrieval made for that point.

Once the placeholders for the variables are defined, I can also assign metadata. The standard units for latitude and longitude are degrees north and east, respectively. AOD is unitless, so none are assigned in it's metadata.

```
latitudes.units = "degrees north"
longitudes.units = "degrees east"
```

The following code combines the three files. Each file is opened and the latitude, longitude, and AOD550 are read in from the file. Because I am combining three files, I define three zero arrays that are three times the size of the first file. I will open each file and iteratively fill with values. In addition to combining the files, I will only save the "best" quality data. So, I will examine the QCAll variable and only include data where this column is equal to 0. Other numbers indicate medium, low quality, or no retrievals. All other data will be set to the fill value.

```
var = np.zeros((rows*num_files, cols))
lats = np.zeros((rows*num_files, cols))
lons = np.zeros((rows*num_files, cols))

for filenum, filename in enumerate(filenames):
    file_id = Dataset(filename, mode='r')

    i1 = rows*filenum
    i2 = rows*(filenum+1)

    # Get coordinates
    lats[i1:i2, 0:cols] = file_id.variables['Latitude'][:,:]
    lons[i1:i2, 0:cols] = file_id.variables['Longitude'][:,:]

    # Create and fill variables
    value = file_id.variables['AOD550'][:,:]

    # Simple binary quality flag
    dqf = file_id.variables[ 'QCAll'][:,:]
    value[ dqf != 0 ] = -999.9

    var[i1:i2, :] = value[:,:]

    file_id.close()
```

I fill the new netCDF file with the combined data.

```
# Populate lat/lon/var
latitudes[:,:] = lats
longitudes[:,:] = lons
variable[:,:] = var
```

And finally, I close the file. Without this last step, the file will not be readable by any external programs.

```
# Close original/modified netCDF file
rootgrp.close()
```

10.5.2. Using Xarray to Create netCDF Files

Following the previous section, you may appreciate the simplicity that xarray offers for saving netCDF files. First, I can open and combine all files using a single line of code. In the netCDF4 method, I needed to create a loop to read the data. As in the previous example, I am stacking the rows to combine the three datasets, which in xarray is done using the `concat_dim='Rows'` and `combine='nested'` keywords. On the third line, I use the `where` command to quality control the data and keep only the "best-quality" data.

```
import xarray as xr

fname = 'data/aod/JRR-AOD*.nc'
aod = xr.open_mfdataset(fname, concat_dim='Rows',
combine='nested')
filtered = aod['AOD550'].where(aod['QCAll'] == 0)
```

With the single line `.to_necdf`, I can save the file.

```
filtered.to_netcdf('JRR-AOD-xarray-combined.nc')
```

The netCDF4 method of saving data are highly verbose, while the xarray code is more compact. Additionally, all the metadata from the source netcdf file are conveniently passed to the final file, so the code is very succinct compared to numerous steps that are necessary using the netCDF4 package method. If you wish to delete some of the metadata, note that you can do so using the del command. For example, you can use the following command to delete the units attribute:

```
del(filtered.attrs['units'])
```

While xarray is more compact, the xarray package uses the netCDF4 package as a dependency. So, a working knowledge of the netcdf4 package is useful for understanding and customizing more advanced features in xarray.

10.5.3. Following Climate and Forecast (CF) Metadata Conventions

For long-term archives, I encourage you to follow netCDF conventions. This facilitates data transparency and will make the data easier to interpret across platforms and over time. While following Climate and Forecast (CF) metadata conventions can be quite in depth, here are some important rules:

Coordinates:

- Make coordinate variables for every dimension when possible.
- Use an existing netCDF Convention for your coordinate variables, especially to identify spatio-temporal coordinates.
- Use the same dimensions when two variables use the same coordinates. However, make a new dimension if two variables have the same shape (perhaps by coincidence, both have 12 vertical dimensions) but use a different coordinate scheme.

Variable grouping:

- All values inside a variable should have the same type (numeric, integer, byte, etc.) and units.
- Attributes should apply to the entire variable.
- Dimensions should be spatial and temporal coordinates, if possible.

Attributes:

- Where possible, add variable attributes for `units`, `long_name` (e.g., a name you would label a plot with), and `_fillValue`. You can also set the `valid_range` in addition or instead of `_fillValue`.
- If an attribute applies to the whole dataset, set it to be a global attribute.
- Encode time as an interval from a reference time, such as `'seconds since 1970-01-01'`
- Packed data values: If you need to conserve storage space, use the `scale_factor` and `add_offset` attributes (Section 3.1.1), so that the packed data can be stored as a byte, short or integer and then read in using:

```
unpacked_data = packed_data * scale_factor +
add_offset
```

10.6. Summary

This chapter provides an overview of some of the available formats that you can encode your data in, such as text, pickle, and netCDF, using common packages. Good archiving habits should not be saved for the end of the project, when you are likely pressured by deadlines, but rather be a continuous component of development. For one, projects can quickly grow bigger than we expect. For example, when making plot titles I recommend extracting the relevant date, variable, and satellite name from the filename or metadata data, not by typing this information into your code by hand. If you reuse the code later, you do not have to manually update the title. Second, deadlines will tempt you to quickly finish up the project, and the data can be cryptic to interpret later. Taking the extra time today to curate your data will save you time in the future. In a community sense, doing so makes it easier to share your results with others.

At this point, you have learned many fundamental coding techniques using Python that are relevant to working with satellite data. In the next section, I will discuss how to integrate Python into a larger workflow. The concept of workflow goes beyond the syntax of how to perform a task to address how we can manage and organize our projects as whole.

Part III

Effective Coding Practices

11

DEVELOPING A WORKFLOW

Cultivating a community-minded a code environment, practice, and style can improve programming efficiency for the scientist and make code easier to share. This chapter will describe some tools to increase productivity and strategies to improve your code, such as scripting, using version control, and virtual environments. The chapter also provides an overview of how to plan and develop code in an agile framework.

Once you have gained fluency in working with Earth observations in Python, you will likely look for ways to make your code more efficient, both in terms of code readability and computational performance. In this chapter, I will show some (1) tools to increase your productivity and (2) strategies to improve your code. This is encompassed by the term **workflow**. In context of this book, here I define workflow as *the common coding environment, practices, and style used to ensure programs and data are reusable and sharable in a scientific community and workplace*. Note the phrase "in a scientific community and workplace." Workflows are not universal, but rather, arise from an individual or a group's needs and experiences. Workflows can also evolve as needs within a community change and new technology becomes available.

From my personal experience, I have found it valuable to strategize before writing or updating code. In the past, I would often write code as fast as possible, get the result I needed, and move on to another task. I did not back-up or document my code and I did not write the code in a general way so that it would be easier to update the next time. My functions were highly interconnected with and dependent on other functions. Over time, my code became increasingly complex and more difficult to update. I adopted better habits after I saw that was

Earth Observation Using Python: A Practical Programming Guide, Special Publications 75, First Edition. Rebekah B. Esmaili.

DOI: 10.1002/9781119606925.ch11

spending more time debugging my existing code than if I started over. As my career progressed, I often received code from colleagues, which at times was decades old. In both cases, It was clear which code was carefully planned and which code was hastily written to meet deadlines. In both cases, I learned that code may exist in a community far longer than the author anticipated.

I hope by sharing my story you are motivated to adopt good coding tools and techniques. Beginning in this chapter, I will touch on more advanced topics such as:

- Creating scripts from notebooks to automate tasks
- Saving changes to your code with version control
- Using virtual environments so that code is reproducible
- Managing your coding projects using agile principles

These techniques are not meant to stifle you, but rather to create a system where you can continue to grow as a programmer and establish good coding habits early on. While your workflow may not be the exact same as mine, many of the tools and concepts covered here are relevant to academic, industry, and government communities.

11.1. Scripting with Python

Jupyter Notebook is an excellent tool when you are new to Python. While you were leaning, I encouraged you to code in notebooks because:

- Inline plots are helpful for quickly viewing your work.
- Code snippets must be complete and fully executable, meaning you must think ahead about the entire purpose of the code block.
- You can document your code inside the notebook and within the code block itself.

However, notebooks also have some drawbacks, especially once your projects increase in complexity:

- They can produce unexpected results if the cells are run out of order.
- Notebooks can be difficult to compare using version control and it is not easy to import one notebook within another.
- Using Jupyter Notebook on a remote server is difficult for novices (see Appendix B.2 for instructions).
- To run scripts, you must launch Jupyter Notebook and either select "run all" from the menu or manually run each cell in the notebook.

For me, the last point is the primary reason I turn to **scripting**. For example, generating 100 different images of Hurricane Harvey might be tedious to run in Jupyter Notebook. Instead, I would create a script, which is a collection of complete code that can be executed automatically to complete a task. To run the code, I would execute my script in the **terminal** (also called a **command line**), which is a text-based application for manipulating files on your computer. After entered, the computer interprets the script commands using the **shell**, which is the interface between the text commands and the operating system. Every operating system has a shell, which include Bourne again shell (bash) on Unix/Linux and cmd or powershell on Windows.

All the examples in this chapter use basic commands in the bash syntax. If you are brand new to the command line or shell scripting, I recommend checking out The Command Line lessons on Ryan's Tutorials website (https://ryanstutorials.net/linuxtutorial/commandline.php).

11.1.1. Creating Scripts Using Text Editors

Python scripts are often written in a **text editor**, which is a program that allows you to edit text while adhering to a standard encoding format, such as UTF-8 (Section 5.1). Text editors are different from **word processors**, which support stylized text (such as special formatting like bold and italic as well as various fonts and size options). Because of this, word processors can introduce unprinted characters that will not be recognized by the Python interpreter when it is run.

Most operating systems come with a basic text editor. These are TextEdit on Mac, Notepad on Windows, or vi/vim on Linux. You can use your native text editor, but I recommend downloading and installing the popular Sublime Text editor (https://www.sublimetext.com/) or my personal favorite, Atom (https://atom.io/). Text editors are designed to write code: they make it easy to navigate code, allow you to find and replace variables, and apply color schemes to Python syntax so that it's more readable. **Integrated Development Environments** (IDE) have many of the same features as text editors but may also allow you to integrate with the command line to run and debug you code. Jupyter Notebook is an example of an IDE that uses the web browser to develop code. I also highly recommend PyCharm (https://www.jetbrains.com/pycharm/). I have a list of both text editors and IDEs in Appendix D.1 if you want to explore more popular options. For this section, I will discuss the steps using a text editor.

Open your text editor of choice and create a new file with a `.py` extension. Let's create a very simple script called `hello_earth.py` to get started. In this script, I rewrite the "Hello Earth" example from Chapter 4:

```
#!/usr/bin/env python3

print("Hello Earth")
```

One difference from Jupyter Notebook is that there is a **shebang** at the top of the file. The shebang tells the operating system which interpreter to use – in this case, python.

11.1.2. Creating Scripts from Jupyter Notebook

One of the simplest ways to create a script is to convert an existing Jupyter Notebook. As an example, I created a notebook named `script_example` that only contains one line of code: `print("Hello Earth")`. I can convert any Jupyter Notebook to a script by going to File → Download as → Python (.py), as shown in Figure 11.1.

This will download a new file (`script_example.py`) to your computer. If you open the file using your text editor, you will see:

```
#!/usr/bin/env python3
# coding: utf-8

# In[1]:

print("Hello Earth")
```

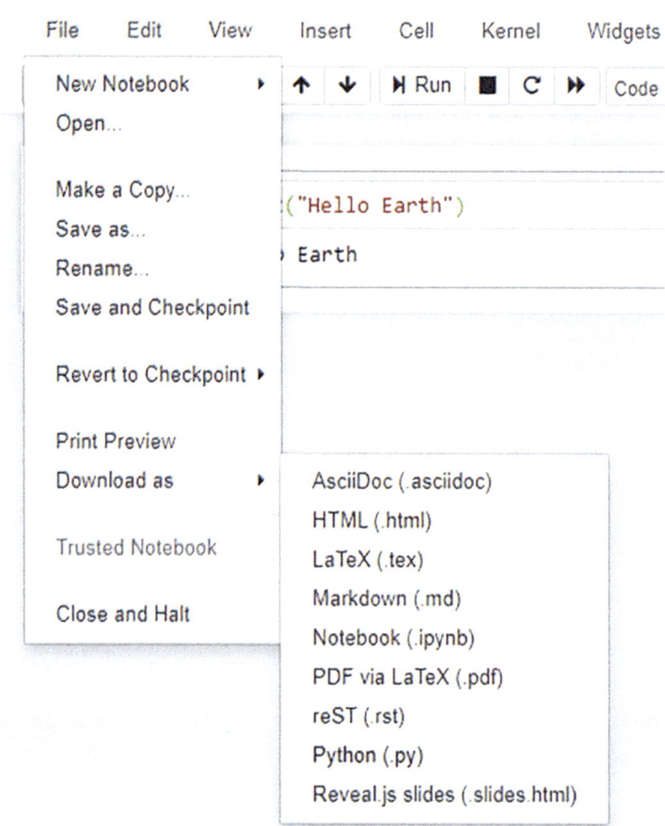

Figure 11.1 Converting Jupyter Notebook to a script.

You will notice that the script contains the line numbers (`ln [1]`), which in my opinion is unnecessary and should be removed from your script. There are ways to turn this behavior off in Jupyter Notebooks, which I show in Appendix B.3.

11.1.3. Running Python Scripts from the Command Line

Now I am finished editing the code and I want to run it. I will discuss two ways you can run Python scripts: one using the command line interpreter and the other using **iPython** (Figure 11.2). iPython is an interactive command line that allows you to run code in chunks. In fact, Jupyter Notebook is built using iPython, which explains the similarity in behavior.

For both options, I will need to open the command line. On Windows, I suggest using the Anaconda Prompt, which you can access from the start menu or using Anaconda Navigator. On Mac or Linux, open the Terminal app. Once the command line is open, you start in a default location. For example, I am using Windows, and when I launch the Anaconda Prompt, I see:

```
(base) C:\Users\rebekah>
```

Now, I want to navigate to where our script is. To do this, I will change directories using `cd`. For example, if your code is stored in `C:\Documents\Python`, you can type:

```
cd C:\Users\rebekah\Documents\Python
```

The command line will now be updated showing:

```
(base) C:\Users\rebekah\Documents\Python>
```

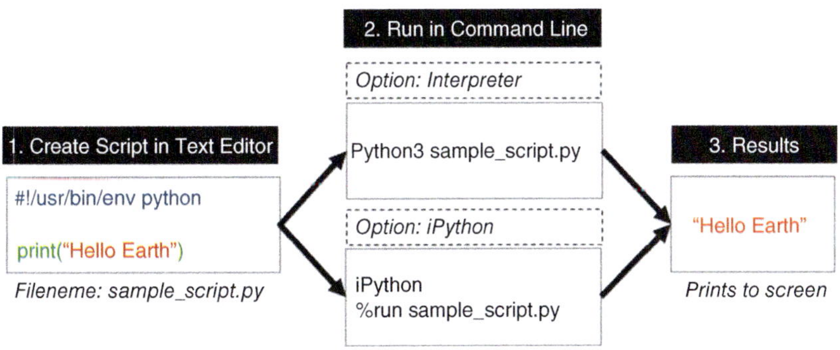

Figure 11.2 A flow chart of the steps necessary to run scripts in the command line.

Now that I am in the right place, I can call the **Python interpreter**, which converts your code into a format that your computer can understand and executes the command. If you installed Anaconda, this includes a Python interpreter than you can use (`python3`). So, to run the script, type:

```
python3 hello_world.py
```

If successful, "Hello Earth" should print to your screen.

A second method is to use `iPython`, which allows you to open Python in interactive mode. Unlike the command line method, iPython will let you run code line-by-line. So, like Jupyter Notebook, you have the option to copy and paste you code from the text editor in chunks into the iPython window. You can also call the entire script inside iPython. This is done by starting iPython and using the command `%run [script name].py`. Below is a capture from my terminal:

```
Python 3.7.6 (default, Jan  8 2020, 20:23:39) [MSC v.1916
64 bit (AMD64)]
Type 'copyright', 'credits' or 'license' for more
information
IPython 7.12.0 -- An enhanced Interactive Python. Type '?'
for help.

In [1]: %run script_example.ipynb
Hello Earth
```

One advantage of using iPython is that after the script finishes running, variables that were generated in the script are still in memory. Then, you can print or operate on the variables to either debug or to develop your code further.

You may have noted two differences in workflow for write code in scripts versus notebooks, (1) that code cannot be inline and (2) the program must run fully to the end. Your code will certainly be more complex than this example. No matter your level of expertise, it is unlikely that you will code your program perfectly without any errors on the first try. To help avoid errors, text editors and IDEs will directly interface with the command line so that you can run your code inside the editor. Furthermore, some text editors and IDEs offer syntax checkers, such as vim flake8, PyCharm, and Sublime text (Section 12.1.2). Another method of debugging is to continue testing code in notebooks, where you can run code line-by-line to check the output, and then export it to the script like I showed in Section 11.1.2.

11.1.4. Handling Output When Scripting

In the previous example, I printed text to the screen, but Python is capable of saving figures and data. Recall from Section 10.1, the plots can be saved to file using the `plt.savefig()` instead of `plt.show()`.

It is possible to directly display your graphics using the X11 protocol (by default in Linux) with XQuartz (Mac) or PuTTy (Windows). I typically discourage this because satellite imagery tends to be very large and thus slow to display remotely. From my experience, it is usually faster to write an image to a file and then view the plot after it is fully rendered. While I wanted to make you aware of this option, I won't describe how to set this up in this book.

11.2. Version Control

Version control is a way of tracking changes to your code over time. When developing code, version control benefits you because you can easily roll back to an older version if you introduce a bug. If you are collaborating on code with others, version control helps keep track of the differences between the multiple versions. I recommend **Git** due to its popularity and because Git easily connects with **online repositories**, which can host your Python projects and documentation online. You can install Git with Xcode Command Line Tools on Mac OS and with Git for Windows (https://gitforwindows.org). Linux users can install Git using a package manager, such as apt, yum, and dnf. While Git is a powerful tool with many features, in this section I will only cover the basics.

First, using the command line (Section 11.3.2), you will initialize a **repository** for your code. This defines and labels which files are being tracked for changes. It is typically the parent directory of the project. In theory, you can add any number and kind of files to your repository. For example, you can add netCDF files to your repository, but it's not likely that you need to track changes in these files. Furthermore, because the files are large relative to the size of your code, tracking changes in netCDF files in your repository will slow down the time it takes to save the changes. So, you want to be selective about what files you add to your repository.

From the command line, change your directory to your project using the `cd` command (e.g. cd `C:/Documents/python_projects`) and type:

```
git init
```

This starts Git in the current directory. If this is your first time running git, you will see a prompt asking you to type your name and email. For practice, I will create a text file with the words "Hello Earth" and name it `git_practice.txt`. Then create another empty file called `irrelevant.txt`.

Next, you will specify which files are tracked using `git add`. All other files are not tracked. For the example, I will add `git_practice.txt`:

```
git add git_practice.txt
```

To check if the file was successfully added to the repository, type:

```
git status
```

Similarly, to remove files from being tracked, use `git remove filename`. For example, if you add `irrelevant.txt`, you can remove it from your repository using `git remove irrelevant.txt`. This does not delete the file from your repository, but Git will no longer track the changes.

If you make small changes and want to save the latest version of the code but not any of the intermediate steps, use the `git add` command again to update the file. Once you make a major change, the code is in a state that you want to make a permanent checkpoint. You can return to at a later time if you `commit` the change. This will give you a specific code version to return to, which includes a hash identification number specifically pointing this version.

To save them to the repository, use the `git commit` command to either update an individual tracked files. The `-m` option will add a commit message that will detail what the changes were.

```
git commit [filename] -m 'initialize the project
repository'
```

Adding commit messages is mandatory in Git. If you do not add the `-m` option, a text editor screen will appear, prompting you to add a message regarding what has been changed in the code. A good commit message has a clear and short subject line but can contain a more details in the body of the message. Stylistically, it's recommended that the subject complete the sentence "If applied, this commit will _____" in 50 characters or less. I will discuss this more in Section 12.1.1.

You can now continue to make changes to the code, and when ready to save your current version of the project, use the `commit` command. There are many useful cheat sheets for git version control available online.

11.2.1. Code Sharing though Online Repositories

Online code repositories are websites that can host code. While many of them are for-profit entities, others offer free services. The benefits of using these commercial services is that the code is not linked to a particular institution. Therefore, if you change your affiliation you can continue to access and control your code. Additionally, these platforms are highly searchable and thus provide access to a broader user community outside of the science community. The online repositories also respect licensing, accessibility, and the spirit of open-source code.

In the next section, I show a short tutorial on setting up the GitHub, but the steps are not significantly different if you decide to use other platforms. Other platforms include Bitbucket and GitLab. Both GitHub and Bitbucket are free to use but are commercially owned by Microsoft Corporation and Atlassian Inc. GitLab is open source, but the set-up process is a little more advanced.

11.2.2. Setting up on GitHub

Follow the steps below on the GitHub website:

- Go to www.github.com and set up an account. Log into your account.
- Create a new online repository: go to your GitHub page and click on the + button and select "Create new project."
- Pick a name for the repository and add an optional short description.

Now, switch back to your command line. GitHub provides the exact code for you to copy and paste into your command line.

If you have new project and have not initialized your project:

```
echo "# test" >> README.md git init
git add README.md
git commit -m "first commit"
```

If you created a repository in the previous section, you can add to it using the following code:

```
git remote add origin https://github.com/[username]/
[repository].git git push -u origin main
```

If successful, you should be able to refresh the webpage and see your code in the online repository. If you have a README.md or LICENSE.txt file in your repository, these will automatically be added to your repository.

When you commit, the changes stay local. Every so often, use `git push origin main` to upload the code online to the repository. This allows you to navigate through previous commits, older versions of the code, and share with other group members. If anything happens on your computer or server, your code will be backed up online. While you can do all of these tasks via the command line, I most often rely on my GitHub page to access my projects and share them with others.

11.3. Virtual Environments

Anaconda comes with over 250 installed packages, and you can import any (or all) of these packages while working in the default `base` environment. However, in practice most projects should only require a small subset of these packages. So instead of working in the `base` environment, I can create a **virtual environment**, which is a self-contained installation of Python that only installs the necessary packages for a specific code or project.

Keeping separate virtual environments between projects is useful because it (1) mitigates the impact that package conflicts has on your anaconda installation (2) ensures that you and your colleagues can re-run the code using exact same packages and versions in the future. Package conflicts arise because packages often depend on other packages, but they may have a mismatch in the dependency package version. If `conda` detects a mismatch, `conda` will ask you to update or downgrade the version, which could cause unexpected behavior if the versions are significantly different. If you encounter significant problems and bugs because of package conflicts, it is a lot easier to deactivate and recreate an environment than to reinstall Anaconda. Furthermore, environments also make it easier to distribute code, because you can communicate the exact packages and versions you used to get your results.

While this book uses Python 3, you may come across code written in Python 2.7. **I strongly discourage you from writing any new code in Python 2.7**, as it is no longer being maintained. However, occasionally there are reasons for working with or updating Python 2.7 code rather than updating to Python 3. In these rare cases, you can create a new environment that installs Python 2.7 in an isolated manner. As you become more sophisticated with your code, I suggest you also learn how to work with different environments.

I will describe how to change environments using `conda`. Keep in mind that that there are a number of other tools to manage your environments (such as pyenv).

Working with environments requires three steps:

1. Create the environment.
2. Activate the environment to enter it.
3. Deactivate to leave the environment.

Step one is by far the most complicated because you must communicate to the new environment which packages you want to install, optionally including their version numbers. Below, I will start with the simplest case and build complexity.

11.3.1. Creating an Environment

If you prefer using a GUI, you can create a new environment inside the Anaconda Navigator. Anaconda provides excellent tutorials on package management (https://docs.anaconda.com/anaconda/navigator/tutorials/manage-packages/), so I encourage you to visit there if you are uncomfortable using the command line.

The same steps can also be completed from the command line, e.g. Anaconda Prompt on Windows, Terminal on Mac/Linux. If you used the Anaconda 3 installer that I recommended in Appendix A.1, the command below will create a new installation of Python 3:

```
conda create --name pyviz
```

The result is a clean environment that I named `pyviz`. When I say clean, I mean that this environment will, by default, not have access to any packages. If, for example, you want to import NumPy in your code, you must explicitly pass a list of packages that want to be installed in this environment. For example, to install NumPy, type:

```
conda create -n pyviz numpy==1.17.2
```

Most likely, you will need to install many packages and it is inconvenient to type a long list. For instance, you may want to install Pandas, netCDF4, scipy, and xarray in addition to NumPy. Fortunately, you can write your complete list of packages and versions out in a text file:

```
netCDF4==1.5.2
numpy==1.17.2
pandas==0.25.1
scipy==1.3.1
xarray==0.13.0
```

Then, instead of writing the individual packages, pass this text file using the `--file` argument:

```
conda create --name pyviz --file requirements.txt
```

Then `conda` will search the channels and install the packages you explicitly specified (Appendix A.2). However, one or more package might not be available in your channels. To see what channels are being searched in conda, type:

```
conda config --show channels
```

Some packages are only available on `conda-forge`. You can add `conda-forge` to the list of channels that `conda` searches when you create an environment using:

```
conda config --append channels conda-forge
```

If you want to further streamline this process, you can create a YAML Ain't Markup Language (YAML) file that includes the environment name, which channels you need to search, and a complete list of your requisite packages and versions. YAML is not a different filetype, but rather a text formatting framework that is used by a number of different coding languages, particularly for configuration files.

You can open your text editor and write out the packages you need. Below is an example of how the file is formatted:

```
name: pyviz
channels:
- conda-forge
- defaults
dependencies:
- python=3.7
- cartopy
- dask
- matplotlib
- netcdf4
- h5py
- numpy
- pandas
- proj4
- xarray
- ipykernel
- notebook
- nb_conda_kernels
```

YAML files are case and space sensitive. The first line names the environment (`name`), which helps to distinguish multiple environments. The list following `channels` provides a list of package repositories to search. The `dependencies` include a list of packages. Note that you must include `notebook` (for Jupyter Notebook) or ipython (`ipykernel`) if you want to use either tool in the new environment. Otherwise, you will only be able to run Python scripts.

The above example or something similar can be saved to a file called `environment.yml`. When you create your environment, point to the YAML file (rather than typing each package) to import all libraries at once:

```
conda env create -f environment.yml
```

11.3.2. Changing Environments from the Command Line

Once you have a complete list of your requisite libraries, activate the environment:

```
conda activate pyviz
```

To see if you were successful, the command below will list all available environments with an asterisk next to the active one:

```
conda info --envs
```

You may also want to save a current set of installed packages used to run your code. To create a list of your current environment's dependencies into a YAML file, type:

```
conda env export > environment.yml
```

To exit your current environment, use the deactivate command. Note that this does not destroy the environment; you can switch back into at a later time if you wish.

```
conda deactivate
```

If you are truly finished with an environment, you can get rid of it completely using `conda remove` (it will not be recoverable, unless you created a YAML file):

```
conda remove --name myenv --all
```

11.3.3. Changing Environments in Jupyter Notebook

After activating the environment (pyviz), when you launch Jupyter Notebook it will use the packages in the new environment and not those in your base environment. However, Jupyter Notebook will not change into the new environment in an existing session, or if you launched it using Anaconda Navigator or a Jupyter Notebook shortcut on your desktop. To switch between environments in an existing Jupyter session, you can install an extension to Jupyter Notebook called `nb_conda`:

```
conda install nb_conda
```

Restart Jupyter Notebook, and when the browser launches, you will see a tab called `conda` (Figure 11.3). Clicking it will reveal any environments you created in the command line and allow you to make new ones using the [+] icon. Note that you will need Jupyter installed in the environment, or it will not be available for you to select.

Then activate the above environment. If you have an existing notebook and you wish to switch environments, use the menus to navigate to the environment, which is shown in Figure 11.4

```
Kernel → Change kernel → [environment name].
```

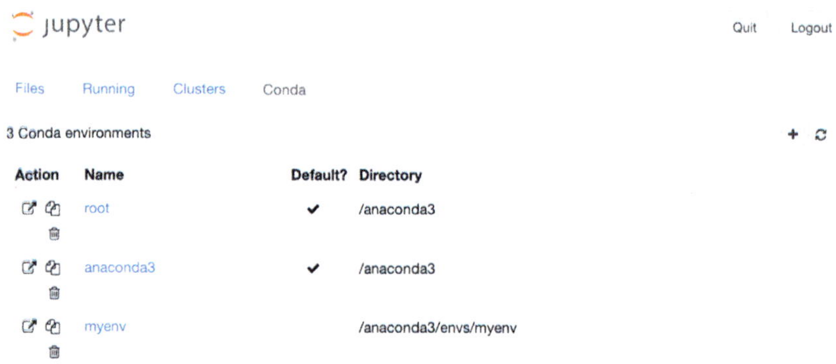

Figure 11.3 Installing nb_conda adds a new tab to Jupyter Notebook, which allows you to switch environments. Source: Project Jupyter.

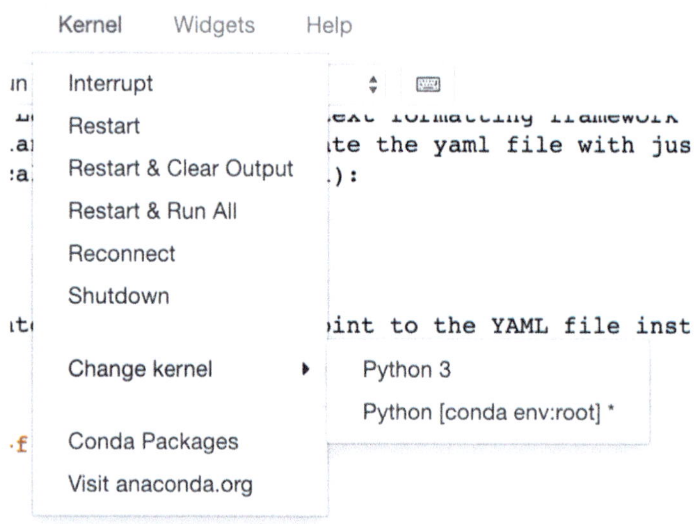

Figure 11.4 Changing kernel inside Juypter Notebooks.

11.4. Methods for Code Development

To improve code quality, scientific programmers can borrow some fundamental techniques from software programmers, such as development methods, requirements and style specification, and debugging. This section provides an overview of a few code development methods you may want to consider.

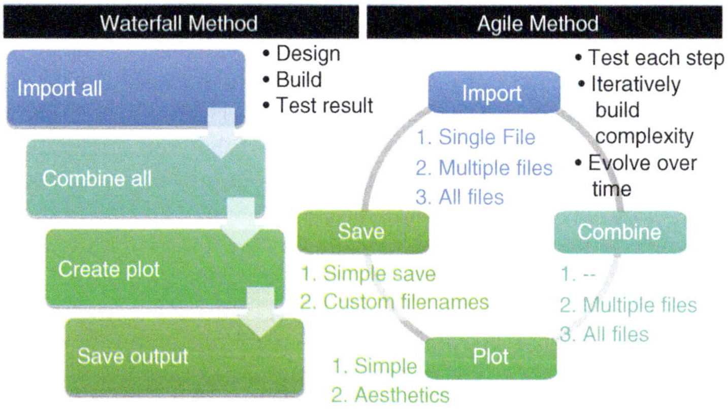

Figure 11.5 Comparison of waterfall and agile framework.

When we begin learning to code, we often employ what is known as a **waterfall method** to design our project (Figure 11.5). Before we start programming, we may plan a rough outline of the coding steps we need to take to complete our project. In the waterfall method, you work on these steps sequentially. For example, let's say you want to create and save an RGB plot (Section 9.4) for a seminar you are giving in a week. You may plan to import ABI channels 2 and 3 (step 1), then regrid channel 2 to match channel 3 (step 2), and combine them using the NDVI formula (step 3) to create plot showing NDVI (step 4), and finish by saving the plot to file (step 5). For small projects, the waterfall method may work fine. After further consideration, you decide to generate an animation showing the NDVI changes over several years instead of single plot. However, in this more complex plan you may have trouble completing one of the upstream step. Since the code is longer, you may have trouble debugging your code. The risk of using the waterfall method is that since each step depends on a prior one, if one fails you will not have any plots for your seminar.

For complex projects, I recommend using the **agile** or **lifecycle** method of code development. The agile method was advanced by Silicon Valley programmers to rapidly build sophisticated software. The agile method breaks down the coding process into smaller chunks that can be solved independently. The agile method also encourages iteration: you add additional complexity to the program each time you work on it. As a first iteration using the earlier example, you could import a single file, create a single NDVI plot, and save it to file with a simple filename, such as NDVI_test.png. In the second iteration, you can improve the aesthetics of the figure by adding a color bar and axes labels, and then the figure save with a detailed filename that includes the date and scan times of the image which are pulled from the metadata in the file. In the third iteration, you might create a loop to plot *t*, adjust the color scales and create a detailed title, and format the positioning of the colorbar.

This method is advantageous if you need to immediately present your results before you are ready, since you will have something available to show. Second, it is easier to debug code if you test it as you go rather than inspecting the result. As mentioned, projects can easily increase in scope over time. You might discover that you do not actually need all those extra details that you originally planned. Additionally, the needs of the consumer of your image may have changed – maybe your advisor or boss wants to examine a different region or add in another dataset. Under the waterfall method, the code quickly become too complex to easily make these changes. Using the agile approach, it may be easier adjust your goals in the middle of the task.

In my discussion above, the agile method is more of a philosophical approach to coding than an actual coding technique. In practice, you will likely use both waterfall and agile for simple and complex projects, respectively. In summary, advanced planning can significantly help with meeting deadlines and ensuring that code is upgradable, rather than hastily written to meet a short deadline.

11.5. Summary

This chapter covered some of the tools that will help you integrate Python into your daily work. You have seen how you can use scripts to automate tasks that you once had to manually run in notebooks and use virtual environments to ensure that the correct packages are installed even when the code is run on a different computer or server. Version control helps track what changes are made to the code over time and to keep copies of older versions of the code in case a new feature breaks the code. Many of these ideas were adopted from the software development community but they are helpful as the scope of our projects increase.

The techniques in this chapter also make it easier to share your code and results. In the next and last chapter of this text, I will discuss how topics such as code techniques, documentation, and licensing can help you share your expertise and results with your research community.

References

5 easy steps to start editing python notebooks over SSH. (2017, November 6). Retrieved April 19, 2020, from https://fizzylogic.nl/2017/11/06/edit-jupyter-notebooks-over-ssh/

Atom: A hackable text editor for the 21st Century. (n.d.). Retrieved November 25, 2020, from https://atom.io/

Git for Windows. (n.d.). Retrieved November 25, 2020, from https://gitforwindows.org/

Linux Tutorial – Master the Command Line. (n.d.). Retrieved November 25, 2020, from https://ryanstutorials.net/linuxtutorial/commandline.php

Managing packages — Anaconda documentation. (n.d.). Retrieved November 25, 2020, from https://docs.anaconda.com/anaconda/navigator/tutorials/manage-packages/

Martin, R. C. (Ed.). (2009). *Clean code: a handbook of agile software craftsmanship*. Upper Saddle River, NJ: Prentice Hall.

PyCharm: the Python IDE for professional developers by JetBrains. (n.d.). Retrieved November 25, 2020, from https://www.jetbrains.com/pycharm/

Sublime Text – A sophisticated text editor for code, markup and prose. (n.d.). Retrieved November 25, 2020, from https://www.sublimetext.com/

12

REPRODUCIBLE AND SHAREABLE SCIENCE

Sharing scientific code reduces barriers in scientific achievement. This chapter describes some techniques to adhere to best coding practices, such as ensuring clear variable and function names and writing modular and reusable code. To write clean code, Python has stylistic rules, which are known as Python Enhancement Proposals (PEP) 8, and tools such as PyLint and flake8 to help both programmers and scientists to follow conventions. Clear commenting and documentation should be practiced for each new program written or when scientist revises existing code.

While access to satellite datasets has grown, the tools to read, analyze, and visualize the data typically have remained hidden behind closed doors. This is a significant barrier to using satellite data and highlights the importance of code sharing to increase the accessibility to scientific data and knowledge. Therefore, the importance of expanding access to easy data portal searches to retrieve satellite data and the software tools used to analyze it cannot be overstated. Important coding principles set forth in this text that include persistent product identifiers, clear metadata, common vocabularies and file structure, and following community standards all contribute to the sharing of scientific knowledge. Adhering to free and open source principles is not limited to data and analysis housed in large agencies, but also applies to individual researchers in industry, government, and academia. If we all follow the guidelines as a community, free and open source science will ultimately increase the productivity and rate of discovery for the whole community.

What practitioners do will ultimately feed back into the improvement of the satellite datasets because often research drives the development of **operational**

Earth Observation Using Python: A Practical Programming Guide, Special Publications 75, First Edition. Rebekah B. Esmaili.

DOI: 10.1002/9781119606925.ch12

datasets that agencies produce; once data are in production, the users themselves provide feedback into the research. By sharing tools with other researchers, datasets are more fully utilized, and satellite data providers can improve existing datasets to address community needs and develop new ones as demand grows.

Skill sharing also includes both human and machine aspects. For scientists, human aspects relate to the accessibility and reusability of the data and code. This means clean and updatable code, clear and concise documentation, and making code available online are important needs. On the machine side, data should be "FAIR" – Findable, Accessible, Intercomparable, and Reusable (Wilkinson et al., 2016). This means that code and data should have a persistent identifier (e.g., a DOI), clear and relevant metadata, follow community standards, and include a usage license.

12.1. Clean Coding Techniques

Always leave the campground cleaner than you found it.

– *Rule from the Boy Scouts of America*

The quote above applies to campgrounds and code alike: if you repeatedly returning to the same code, make an improvement to it each time. This does not require you to perfect it, but this philosophy will allow you to make progress over time.

In the flip side, increasing clutter in your code leads to **technical debt**, where the code becomes increasingly complicated and difficult to update. Eventually, you must either invest significant time rewriting the code to reduce the complexity. In the worst case, you may have to abandon the code and start anew (in my analogy, you can think of this as bankruptcy).

Considering the "cost" of poor coding practices is appropriate because there can be actual economic and productivity loss because of technical debt. For example, Twitter was originally developed using an interpreted language called Ruby on Rails, which made it easier to develop code and update it quickly but at the expense of computational speed. As Twitter became more popular, the search performance slowed significantly because of this initial design choice. Twitter then had to make an expensive investment to rewrite its platform in Java to meet customer demand. If you take the time now to write good, clean code, you will save yourself time (and possibly money) in the future.

Table 12.1 shows some example of how bad habits can make code unusable, particularly in scientific code (SourceMaking.com has some great descriptions and interactive examples of bad coding practices https://sourcemaking.com/refactoring/smells):

Table 12.1 Examples of Code Problems and Solutions

Problem	Definition	Solution
Bloat	An accumulation of functions, methods, and dependencies that become so long and require so many parameters that they are difficult to work with.	Write functions that do one action well, which will limit the number of requires parameters.
Change prevention	Your code is difficult to update because a change in one function or part of your code requires many fixes throughout the program. This high level of interdependency is known **tight coupling**.	Write functions and code blocks that are general and independent of each other.
Cruft	Pointless, redundant, or unused code or comments in your program.	Use version control to track changes and delete obsolete code.

According to Robert C. Martin's canonical Clean Code textbook, "Clean code does what it's supposed to do without breaking. It is easy to read, maintain, and extend. It functions without problems or defects and is written so that it's easy for the next person to work with." Below, I will highlight some beginner friendly principles for developing code:

- **Variable names** should be nouns, not verbs. Avoid using single letters or generic words like *data* or *array*. At the same time, they should not contain too much detail. In the code block, define variables within sections where they are used. This contrasts with legacy programming languages like Fortran, that requires all variables to be defined at the beginning of the entire code block.
- **Function names** should include verbs, such as make, write, open, calculate, etc. For example, the pretend function, *open_SST(filename)*, has a good, descriptive name for its function of opening a file and returning SST data. Functions should be designed to do one thing well – so if you find yourself needing two verbs for the same function, it may be better to break the function into two parts. For example, `open_and_regrid_SST()` could be split into smaller, more digestible functions, `open_SST()` and `regrid_SST()`. When defining functions, it is good to keep similar and dependent functions together in your code.
- **Aesthetics**, such as how you group and space lines within your code, make your program easier to update. Group together the lines of code that illustrate a concept or are related. Conversely, separate different concepts with a new line. Every line of code should be short and digestible.

Things to avoid:

- Rigid code that is difficult to change
- Fragile code that is susceptible to bugs and crashes
- Code that is not reusable for other tasks
- Repetition in the code
- Overly complex code
- Code that is hard to understand

In practice, it is unlikely you will write clean and efficient code every single time you set out to write a program. Seeking perfection is stifling. This is where the agile method can help: on my first iteration, my code is simple and manageable while performing the intended task. I then add incremental complexity with each iteration. As my code becomes more complex, I also perform **refactoring**, which is when we restructure code to improve in both the function and cleanliness of the code. Each time I return to my code "campground," I consider the principles in this section and ask myself if I can improve my code.

12.1.1. Stylistic Conventions

Python enhancement proposals (PEPs) consist primarily of documentation that explains the rational for new features, but also provides guidance to the community on how to stylize code. Of these, the stylistic rules in PEP 8 are the ones that the Python community follows closely. The clean coding principles apply, but PEP 8 provides added details such as using underscores to show spaces in variable names, all caps to show constants, and camel case to indicate a class. These style suggestions are aesthetic but also adds to code consistency.

12.1.2. Tools for Clean Code

Since there are many stylistic rules, it can be difficult to remember all of them, but there are several tools to perform **linting**, or ensuring that the PEP 8 style is followed. Below are two of these tools you can install:

- **PyLint**
- **flake8**

Both tools perform stylistic linting, where it checks to ensure the code follows the defined PEP eight conventions. Additionally, these tools can perform logical linting, which checks for code errors, code with unintended consequences, or patterns in your code that can lead to crashes.

12.2. Documentation

When you find yourself repeatedly using and updating the same piece of code, it means that your code is a solution to some problem. Thus, other researchers can benefit from your code and there will be value in sharing it. This requires properly documenting your code before distributing it to save other code users time and energy and to make your code transparent and understandable. Occasionally, we ourselves are the end user – I have many times returned to old code only to find that I have forgotten how to use it. There are two methods of documenting your code: external to the code using a README file, or inside the code through commenting. Different workplaces and communities (both scientific and open-source projects) will have their own individual cultural expectations for what should be documented inside a code block. This section describes some of the standard documentation practices that are universal to the Python community.

12.2.1. Comments and Docstrings

One way of documenting code is using **document strings** ("docstrings"), which begin and end with quotes (""") or ('''). The first statement in all functions, classes, and modules should include a docstring stating at minimum what the code does. A single line docstring is enough for simple functions. For more complex functions, the docstring should additionally contain the input arguments and the values returned (PEP257; Slatkin, 2015). The closing quotes for the docstring should be alone on the final line.

Comments occur within the body of code or functions. Ideally, comments should only be used to *clarify* code or *warn of consequences* (Martin, 2009). For science code, this may also cite the source of an empirical equation or constants. Python is a "readable" language (PEP20), and the comments should in turn also be readable – use complete sentences and uses proper capitalization and punctuation.

Inline comments, which are comments on the same line as the code, are generally discouraged:

```
List =  ["Himawari", "GOES", "Meteosat"]  # Geostationary
satellites
```

In the scientific community, function arguments, outputs, and dependencies are often included at the top of the function definition. For example, for the haversine function in Section 9.5.1:

```
def haversine(deglat1, deglon1, deglat2, deglon2):
"""Calculates the distance in km between two geolocated
points.

Arguments:
     deglat1, deglon1: the latitude and longitude
coordinate point 1
     deglat2, deglon2: the latitude and longitude
coordinate point 2
```

```
Returns:
    A float of the distance in km between the two points.
"""
    lat1=deglat1*np.pi/180.0
    lat2=deglat2*np.pi/180.0

    long1=deglon1*np.pi/180.0
    long2=deglon2*np.pi/180.0

    # The code below uses the haversine formula.
    a = np.sin(0.5 * (lat2 - lat1))
    b = np.sin(0.5 * (long2 - long1))

    # Diameter of the Earth is 12742 km.
   dist = 12742.0 * np.arcsin(np.sqrt(a * a + np.cos(lat1)
* np.cos(lat2) * b * b))

    return dist
```

Too many comments are also not helpful. As an example, the comment below is redundant and should be removed since the code command clearly explains what is happening:

```
# Imports packages
import numpy as np
```

I discourage commenting out unused code because that makes the code difficult to read. Instead, use version control (Section 11.2) to track changes to the code over time.

Lastly, comments that contradict or are outdated compare to the code below can confuse future users of the code. So, update all comments along with the code it describes.

12.2.2. README File

When I receive a new piece of code, I begin by first understanding how to run it. Therefore, a well-written README file will enhance the functionality of your code with a clear explanation of how it is run. README files contain information on how to install and run code or software and also the intended goals.

While there is no standard structure, I suggest that a good README file should include the following sections, where applicable:

- Title: A detailed name for your project.
- Getting started: information on where to access the program, any relevant background literature, and external software dependencies.
- Installation process: You can explicitly assume some basic familiarity with software tools (e.g., that they have Python) on the user's part. If there are any external tools, then mention them here.
- Required libraries: You can explicitly list them or point to any requirements or environment files that are included in the distribution.
- How to use or run the code: This is a helpful guide for others using your code, or for yourself when returning to the code at another time.
- Author(s): You and any other contributors to the project.
- Resources or references: Any materials used to develop the code or further resources on your project.

There is no definitive template for a readme file; however, the above is a good start. Through this book, you should have collected a portfolio of practice projects. Practice creating a README file for one of these.

12.2.3. Creating Useful Commit Messages

As mentioned in Section 11.2, commit message are mandatory in Git, and it is worth putting some thought into these messages. These commits provide a history for changes to the code. Over time, the code will become increasingly complex and you will not be able to visually find the differences. A good commit message should help the user understand what changes were made to the code without having to compare versions of the code.

A useful commit message has these features:

- It will complete the following sentence: Applying this commit will . . .
- It is less than 50 characters; longer messages will be cutoff on most websites.
- Minor changes or additional details are in the body of the commit message.

The **SciPy** community (which develops **NumPy, Matplotlib,** and **Pandas**) uses additional recommendations for writing good commit recommendations. The messages should help the user understand what changes were made to the code without having to compare versions of the code. They also recommend starting a commit using a standard set of acronyms, such as:

```
BUG: bug fix
DEP: deprecate something, or remove a deprecated object
DEV: development tool or utility
DOC: documentation
ENH: enhancement
MAINT: maintenance commit (refactoring, typos, etc.)
REV: revert an earlier commit
STY: style fix (whitespace, PEP8)
TST: addition or modification of tests
```

words can communicate ideas and statements. Much like the layered approach described above, the process of designing a figure involves three stages:

1. **Specification.** The components (or objects) that go into the plot. This includes the data, any data transformations, the choice of map projection, aesthetic attributes, legends, and color schemes. These are all the necessary components to make a visual.
2. **Assembly.** The actual construction of the elements in the specification. Before aggregation, the items are standalone objects and do not inherit any properties from their parents.
3. **Display.** The actual rendering following the assembly of the objects.

The above stages also permit easy revision. Changing the data from one plot type to another allows us to communicate different ideas about the underlying data.

12.4.2. Undergo Revision

It is easier than ever to create visualizations on computers (Section 2.1). Figures can quickly be rendered and then refined. In *Eloquent Science* (2009), David Schultz defines the steps that take a rough draft figure to a final one:

1. **Design** as in the type and layout of the plot.
2. **Size** of the final figure; vectorized graphics are easily scaled but may be very large for complex figures.
3. **Aesthetics** that make the figure compact and self-contained.
4. **Consistency** within and between other figures.
5. **Annotation** both within the figure and in the captions.

Creating a plot that satisfies these conditions produces a chart, perhaps even a visually pleasing one. However, in the next section, I will describe how to create meaningful plots.

12.4.3. Are Accessible and Ethical

Selecting appropriate labels, fonts, and colors for a plot is not trivial. Labels and fonts should be easy to read, and the figure should be self-describing through text and graphical annotations. However, choosing colors requires you to consider what information you want to emphasize and make it easier for the reader to interpret. Rainbow is a common color scheme in Earth sciences, but I generally do not recommend using it. The reason is that the high-perception colors create artificial patterns of information. Additionally, the human eye cannot accurately judge the magnitudes that are rainbow colored (Borland and Taylor, 2007). An example of the same plot of wildfire smoke plumes with two different color schemes can be seen in Figure 12.1. The left image uses a two-tone color scheme (purple-blue, or "PuBU"), while the image on the right uses many with sharp transition points (NCL's default color). While the right image is useful for highlighting the smoke

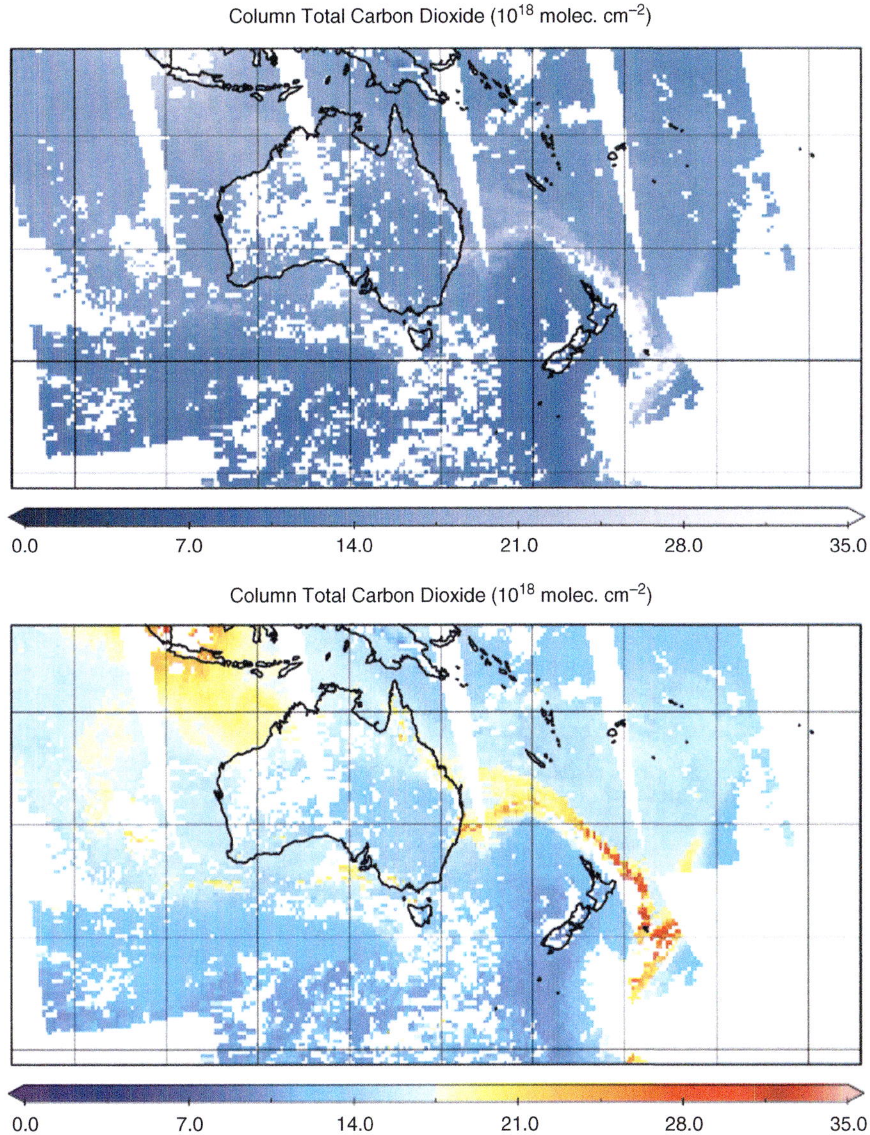

Figure 12.1 Example of uniform color scheme (top) and how a different color scheme causes artificial enhancement of features (bottom) for the same image. Even though the data are using the same range (0, 35.0), the image on the right shows more structure. Both images show total column carbon dioxide (10^{18} molecules/cm^2) from intense Australian wildfires during the afternoon orbit of Suomi NPP on 11 November 2019, from the NUCAPS algorithm. White colors are missing data caused by uniform clouds, precipitation, or gaps in the overpass. Based on Smith et al., 2020.

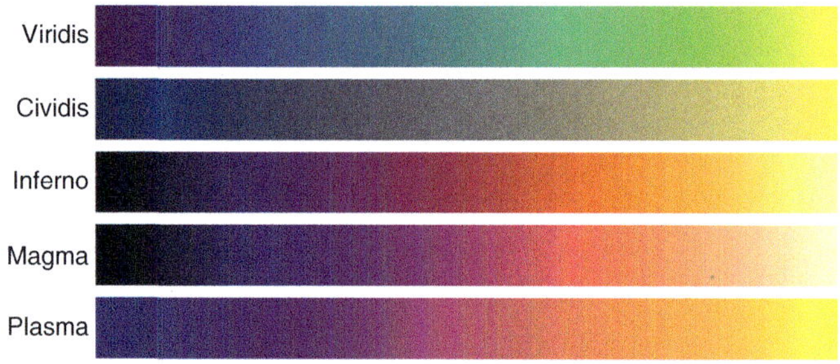

Figure 12.2 Perceptually uniform sequential colors available in matplotlib.

plume, you should be careful not to misrepresent results by using unnecessary high contrast to show patterns where they do not exist.

Some color schemes are useful for quickly understanding certain types of data and analysis. As a result, some conventions have been informally adopted by research communities. Temperature anomaly data shows values that are above and below averages usually use blue for negative values, white for values near zero, and red for positive values. Alternatively, low and high rainfall plots often use brown (low values) and green (high values). There are no standard rules for choosing colors but consider what you are communicating to your audience (Rheingans, 2000).

In addition to the content, when you are choosing colors you should also consider human visual perception. Roughly 8% of males and 0.5% of females experience color blindness, the most common type is insensitivity to red and green (which is called *deuteranopia*). As a result, color schemes should avoid these two colors. Another visual challenge is contrast sensitivity, which peaks at age 20 and declines with age. High-contrast color schemes make it easier for older age groups to distinguish lines. However, as mentioned, some high-contrast color schemes can misrepresent the data. Perceptively uniform colormaps (Figure 12.2) address many of these (Kovesi, 2015). Most perceptively uniform colormaps are color-blind friendly and high contrast, and many are grayscale printing friendly. While not a vision impairment, printing scientific papers and related figures is still a common practice around the world.

When designing figures, no one color scheme satisfies all requirements. Figures should be both truthful and inclusive when presenting information. Often, picking good colors involves trial and error, but there are some useful resources, such as Color Brewer 2.0 (https://colorbrewer2.org/). Also, remember that colors are not the only way to distinguish plot elements. For instance, patterns and line styles (e.g., solid, dotted, dashed) are useful choices for making figures more accessible (and printer friendly). Additionally, you can also choose to spread out information in two simpler plots, rather than one more complex plot.

12.5. Summary

In this chapter, I provided an overview of some of the clean coding techniques and how to effectively document and license your code. When writing and documenting your code, I recommend that you consider the potential users within your research scientific community. This means the code should be clearly written and easy to update, and the supporting documentation should be descriptive. You should also consider the audience when making visualizations, such as using color-blind-friendly colors and large text labels. Many scientists improve their coding skills through hands-on experience, through mentorship by our colleagues, or self-study. By providing your peers with helpful tools and resources, you can contribute to your community in a way that goes beyond to your scientific contribution.

References

Barnes, N. (2010). Publish your computer code: it is good enough. *Nature, 467*(7317), 753–753. https://doi.org/10.1038/467753a

Borland, D., and Taylor Ii, R. M. (2007). Rainbow Color Map (Still) Considered Harmful. *IEEE Computer Graphics and Applications, 27*(2), 14–17. https://doi.org/10.1109/MCG.2007.323435

David, C. H., Famiglietti, J. S., Yang, Z.-L., Habets, F., & Maidment, D. R. (2016). A decade of RAPID-Reflections on the development of an open source geoscience code: A Decade of RAPID. *Earth and Space Science, 3*(5), 226–244. https://doi.org/10.1002/2015EA000142

Design patterns and refactoring. (n.d.). Retrieved November 25, 2020, from https://sourcemaking.com

Design patterns. (n.d.). Retrieved April 17, 2020, from https://refactoring.guru/design-patterns

Gil, Y., David, C. H., Demir, I., Essawy, B. T., Fulweiler, R. W., Goodall, J. L., et al. (2016). Toward the geoscience paper of the future: Best practices for documenting and sharing research from data to software to provenance. *Earth and Space Science, 3*(10), 388–415. https://doi.org/10.1002/2015EA000136

Morin, A., Urban, J., and Sliz, P. (2012). A quick guide to software licensing for the scientist-programmer. *PLOS Computational Biology, 8*(7), e1002598. https://doi.org/10.1371/journal.pcbi.1002598

Peng, R. D. (2011). Reproducible research in computational science. *Science, 334*(6060), 1226–1227. https://doi.org/10.1126/science.1213847

Schultz, David M. (2009). *Eloquent Science: A Practical Guide to Becoming a Better Writer, Speaker, and Atmospheric Scientist*, American Meteorological Society.

Shamir, L., Wallin, J. F., Allen, A., Berriman, B., Teuben, P., Nemiroff, R. J., et al. (2013). Practices in source code sharing in astrophysics. *Astronomy and Computing, 1*, 54–58. https://doi.org/10.1016/j.ascom.2013.04.001

Wilkinson, M. D., Dumontier, M., Aalbersberg, Ij. J., Appleton, G., Axton, M., Baak, A., et al. (2016). The FAIR guiding principles for scientific data management and stewardship. *Scientific Data, 3*(1), 160018. https://doi.org/10.1038/sdata.2016.18

Appendix A

INSTALLING PYTHON

You will need an internet connection to obtain Python and download the tutorials. If you do not have a persistent internet connection, I suggest following the entire setup tutorial below. Then, the remainder of the book can be completed offline.

If you are completely new to Python, you can also run the notes interactively online using Binder, which can be found on the book GitHub page (https://github.com/resmaili/EarthSciViz). Binder sessions create an executable environment in your web browser to run the code in this book. However, once you wish to develop your own code, you will need to install Python on your workstation.

A.1. Download Tutorials for This Book

All tutorials shown in this book, including both code and datasets, are accessible through https://resmaili.github.io/Earth-Obs-Py. If you are unfamiliar with using git and GitHub, the tutorials can be downloaded to your computer by selecting Code ➔ Download ZIP (shown below).

Earth Observation Using Python: A Practical Programming Guide, Special Publications 75, First Edition. Rebekah B. Esmaili.

DOI: 10.1002/9781119606925.appendixA

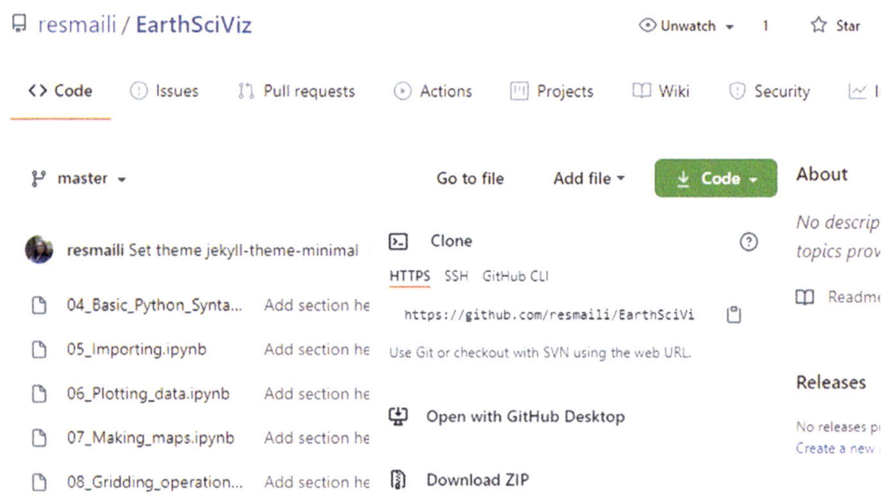

A.2. Download and Install Anaconda

Python is included on many Unix installations. However, I recommend that beginners install the Anaconda package manager. The Anaconda distribution includes standard Python, many scientific packages, and the Jupyter Notebook development environment that I use for examples in this book. Anaconda installs over 150 packages and is quite large (2 GB). If download speeds and space are a limitation for you, you can download Miniconda (400 MB), which contains fewer installed packages.

First, install the latest version of Anaconda (Python version 3.8 or newer, not version 2) appropriate to your operating system, www.anaconda.com/download/ This step is operating-system independent. Miniconda can be downloaded from https://docs.conda.io/en/latest/miniconda.html.

A.3. Package Management in Anaconda

Downloading and installing Anaconda provides you with many scientific **packages**, along with **development environments** such as Jupyter Notebook, iPython, and Spyder. However, you will need additional third-party packages to be able to import, analyze, and plot satellite datasets that I show in this text. You can download and install packages from package **repository channels**, which are like online catalogs of python packages.

If you try to import a package that is not installed, Python returns a `ModuleNotFoundError`, which looks like:

```
ModuleNotFoundError
Traceback (most recent call last)
<ipython-input-1-d16bc6fdb2bf> in <module>
----> 1 import cartopy
ModuleNotFoundError: No module named 'cartopy'
```

I show the steps on how to do this via the command line using `conda` (https://conda.io/en/latest/), an open-source package and environment (Section 11.3) management system. If you are unfamiliar with the command line, Anaconda comes with Anaconda Navigator (https://docs.anaconda.com/anaconda/navigator/tutorials/manage-packages/). Anaconda Navigator is a desktop graphical user interface (GUI) which may be more intuitive to new programmers. For more information on using it to install packages, see Anaconda Navigator's online user guide.

To install packages, you will need to access the terminal on Mac or Linux and the Anaconda Prompt on Windows.

```
conda install [package name]
```

This will check the default conda channel (https://repo.anaconda.com/pkgs/) for the package you specified above. However, the default channel does not contain all possible packages. `conda-forge` is another community channel that contains many libraries and more up-to-date versions than those on the default channel. Many of the packages used in this book are only available on `conda-forge`. You can add `conda-forge` to the list of channels to search using:

```
conda config --append channels conda-forge
```

Since you have several packages to install to work through this book, installing them individually takes time. So instead, you can install all the packages by either downloading or creating the `pyviz.yml` file (https://github.com/resmaili/EarthSciViz):

```
conda install --file pyviz.yml
```

If successful, you shouldn't experience any `ModuleNotFoundError` messages while working through this book.

Appendix B

JUPYTER NOTEBOOK

B.1. Running on a Local Machine (New Coders)

Both Windows and Unix users can open the Anaconda Navigator to launch Jupyter Notebook. I prefer to open the terminal on Mac or Linux Machines or open the Anaconda Prompt in Windows and type:

```
jupyter notebook
```

Windows users can navigate to a Jupyter Notebook icon from the start menu. If it succeeds, a browser window will open and you will see the Jupyter Notebook main page. To create a new notebook, click on New --> Python 3. A new tab will open with an empty notebook. Here, the tab is opened in Mozilla Firefox.

Earth Observation Using Python: A Practical Programming Guide, Special Publications 75, First Edition. Rebekah B. Esmaili.

DOI: 10.1002/9781119606925.appendixB

In the annotated screen capture from Jupyter below, **A** shows the menu options. File contains tasks such as save/rename; edit allows you to add, copy, duplicate, and cut cells. Cell permits you to run code blocks within cells and set the cell type. The latter is useful for converting your notebooks to latex or html. The Kernel menu allows you to stop a command or the entire session if something goes wrong (e.g. you try to open too many files and your computer runs out of memory).

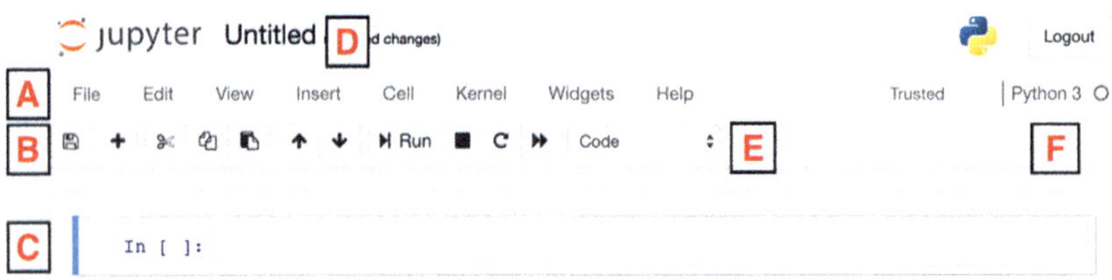

B shows many of the commonly accessed menu commands as icon shortcuts. The icons, from left to right, are save, new cell, cut a cell, copy a cell, duplicate cell, move cell up, move cell down, run active cell, stop the cell, restart the kernel, and restart and run all.

C is where we type code. Once you type your code into the cell, you can click the run button or hit Shift and Enter to run from the keyboard. The output will display below.

In **D**, you can rename the current notebook. By default, it will be called Untitled; get in the habit of giving your notebooks descriptive names because it is easy to accumulate many notebooks over time.

E shows a dropdown menu for you to select cell type. The options include code (default), markdown, or raw. Markdown is a plain text formatting syntax, so it's valuable for documenting your code. Markdown's presence in Jupyter Notebook is part of the reason they are called notebooks; you can mix code and detailed notes and documentation. Raw cell types are just plain text; all spaces and tabs will be preserved in this format.

Finally, **F** shows the status of the kernel. While most notebooks are written in Python, Jupyter Notebook also supports over 100 programming languages, although only include Python by default. So, the top right corner indicates what language the kernel is in and the small dot to the right indicates the status. If code is running, the dot will be filled in and you must wait for the cell to finish running before another is processed.

You are all set up! So, return to Chapter 4 and go through the "Hello Earth" example to test out the above functions.

B.2. Running on a Remote Server (Advanced)

Many researchers work off remote Linux servers hosted by their employer or university, which are often command line only. Below I briefly outline how to access a remote instance of Jupyter Notebook on a workplace server. Note that I assume that you have a working knowledge of the command line and are familiar with how to use the Secure SHell (SSH) protocol to connect your local computer to a remote server. I also assume you know how to use a VPN (Virtual Private Network) client if it is required to connect to your server remotely. If you are not familiar with how to use VPN and SSH, you will need to check with your IT department about how to get set up.

There are two potential ways to run Jupyter Notebook remotely. First, you can launch a browser on your remote server (e.g., Mozilla Firefox) to run Jupyter Notebook. This means you enabled X11 Forwarding set-up and log-in using the `ssh -X -Y` flags. However, I do not recommend this work environment because I have experienced significant lags when displaying the web browser window, which is frustrating. The second method is using your own computer's *local browser* to eliminate this lag. You will need to have Anaconda installed on your remote machine. I personally recommend installing your own copy of Anaconda – for instance, your home directory – rather than use a shared version. However, if your institution has other recommendations, follow their guidelines.

Before I proceed further, I will need to select a local **port number** and determine the remote server's **hostname**. Port numbers are used to identify the endpoints in a line of communication between two machines. The value of the port number is somewhat arbitrary, although certain port numbers are reserved for certain tasks. Typically, 8000 or 8080 are usually available. However, you may need to confirm this with your IT department.

The hostname is a label used to identify which remote server you are connecting to. On the remote server command line, the server hostname address can be found using:

```
hostname -I
```

Next, I need to set up an **SSH tunnel**, which connects content between your remote server and local browser. Using the hostname, type the following command to enable Jupyter Notebook:

```
jupyter notebook --no-browser --port=[port number] --ip=
[hostname]
```

To access the Jupyter Notebook, you can also copy and paste the URL that is generated following from the previous command.

```
To access the notebook, open this file in a browser:
file:///C:/Users/rebekah/AppData/Roaming/jupyter/runtime/
nbserver-18284-open.html

Or copy and paste one of these URLs: http://hostname:8080/
?token=e5f68voes8wcz7bs7m0ssedouziikvk9z41liz6u
```

You can copy and paste this address into your local browser (e.g., Firefox, Chrome, Safari) and access notebooks on the remote server.

B.3. Tips for Advanced Users

If you are comfortable working on a Linux machine, some of the following suggestions may be useful for you.

Use an alias to establish the remote tunnel. You can save the above tunnel command as an alias in your `.bashrc` or `.bash_profile` file. For example:

```
jn=jupyter notebook --no-browser --port=[port number] --ip=
[hostname]
export jn
```

Use a windows manager to maintain a semi-permanent tunnel. Windows managers like `Screen` (www.gnu.org/software/screen/) or `tmux` (https://github.com/tmux/tmux/wiki) are useful in their own right, but you can also use them to create a session where you run your tunnel command (e.g., the one I saved to the alias `jn`), and then detach the session. By doing this, your notebooks will continue to run in the background even if your current connection has dropped or timed out.

If you enable passwords (see the next section on configuration files), you do not need to type the hash each time you run Jupyter Notebook. You can also then bookmark the notebook location in your web browser.

B.3.1. Customizing Notebooks with Configuration Files

Jupyter options customizable inside the configuration file (`jupyter_notebook_config.py`). The configuration file is in your home directory (`~/.jupyter` on Linux/Mac, or the `%APPDATA%` folder on Windows), unless you choose to install Anaconda in another location on your computer. By default, the configuration file will not exist. To generate the configuration file, open the terminal (Linux/Mac) or the Anaconda prompt (Windows) and type:

```
jupyter notebook --generate-config
```

If you check, in your Jupyter directory you will see `jupyter_notebook_-config.py`. Open `jupyter_notebook_config.py` with a text editor.

Some useful customizations include:

- **Setting a password**

For security, the notebook URL will include a very long hash token, which, for example, looks like:

```
    http://hostname:8080/?
token=e5f68voes8wcz7bs7m0ssedouziikvk9z41liz6u
```

If you leave off the token, you will be prompted for a password, which you will need to set up. The advantage of setting up a password is that every time you start a remote session of a notebook, the hash will change. By setting up a password, the location will stay the same (and you can bookmark it in your browser).

Uncomment the line by deleting a # symbol at the very beginning of the line `c.NotebookApp.allow_password_change` (see example below).

```
# While logging in with a token, the notebook server UI
will give the opportunity
# to the user to enter a new password at the same
time that will
# replace the token login mechanism. This can be set to
false to prevent changing password from the UI/API.
c.NotebookApp.allow_password_change = True
```

Then in the command line, set the password by typing:

```
jupyter notebook password
```

You will be prompted to enter and re-enter a password, which will generate a password in `jupyter_notebook_config.json.` Note that it is not encrypted, so for security, do not use a sensitive password. Then, you will not need the hash to access the Jupyter Notebook:

```
http://hostname:8080/
```

Then, you will be prompted to enter a password.

- **Suppressing Script numbers in Jupyter Notebook**

For advanced users, you can suppress this output by customizing your `jupyter_nbconvert_config.py` file (https://nbconvert.readthedocs.io/en/latest/customizing.html).

B.3.2. Starting and Ending Python Scripts

When writing a Python script, the first two lines of code are **encoding declarations,** which are used by the system before processing the code. The first line is called a **shebang**, which tells your system which interpreter to use when running the code. In this example below, I want to call python3 to run my code. This will line will call `python3` based on your `PATH` environment variables, which are set in the `.bash_profile` or `.bashrc` at startup. If your `PATH` points to your local Anaconda Python installation, this version will run instead of the version that is installed on your system. In the second encoding declaration line, there is a statement referring to the text encoding, which is UTF-8. UTF-8 is the default, but you have the option of using other types.

```
#!/usr/bin/env python3
# coding: utf-8

import sys
print("Hello Earth")
sys.exit()
```

In the body of the code block above, I import the `sys` package so that I can call `sys.exit()` to end the script. Explicitly telling the script to end is a recommended practice, but it is not required because Python interpreter will automatically terminate the script when it reaches the last line. In some script examples online, you may see `quit()` at the end of the script. Both functions send the SystemExit exception, which tells the Python interpreter to exit. The `quit()` command comes from the site package, which is imported during startup in most interactive python environments, like iPython. However, not all computer systems will import site automatically, whereas all python installations come with the **sys** package. The Sys package interfaces your python interpreter with functions on your system. You can avoid errors by using `sys` and end scripts with `sys.exit()` and not `quit()`.

B.3.3. Creating Git Commit Templates

I found it helpful to set up a template to help remember some of the standards for writing a useful commit message in git. This template pops up every time I enter my commit message. To create a template, create a file called `.gitmessage` and open it with your text editor. Inside the file, for example, you can type:

```
# Subject: "Applying this commit will ..." (<50 characters)
# What the change to the code is and why the change is made
```

Like in Python, the hashtags (#) are comments and the trailing text will not be saved in the final version of the commit message. However, they are useful for reminding you of the guidelines.

After creating this file, you need to use `git config` to point to it to our template file. If I saved it in my home directory (e.g. `/home/rebekah`), the command becomes:

```
git config --global commit.template /home/rebekah/.
gitmessage
```

Now, each time you type `git commit [filename]`, you will see the above template instead of a black text document.

Appendix C

ADDITIONAL LEARNING RESOURCES

Software Carpentry (https://software-carpentry.org/about/) and Data Carpentry (https://datacarpentry.org/about/) offer excellent online tutorials on Python.

The following two books can be helpful as a supplement to this one. They both provide a more in-depth explanation of Python programming.

- DeCaria, A. J. and Petty, G.W. (2021). Python programming and visualization for scientists, 2nd ed. Madison, Wisconsin: Sundog Publishing.
- Lin, J. W.-B. (2012). A hands-on introduction to using Python in the atmospheric and oceanic sciences.

In terms of conferences, the *Symposium on Advances in Modeling and Analysis Using Python* at the American Meteorological Society is an excellent way of learning about advances in Python related to the atmospheric sciences. While not specific to Earth sciences, the SciPy conference (e.g., https://www.scipy2020.scipy.org/) is an opportunity to learn and also share your applications to a more general audience.

There are several online communities that you can join. You may not see yourself as a Python developer at this point, but contributing to either of these communities can help you rapidly increase your skills while benefiting the broader scientific community:

- **Pangeo project** (http://pangeo.io/). Pangeo is a community that supports an ecosystem of packages that acts as an intersection between science, software, and computer infrastructure. It promotes open, reproducible, and scalable science. The development community provides documentation, develops and maintains software, and deploys computing infrastructure to make scientific research and programming easier. Many of the tools I used in this book are members of its community: **xarray**, **iris**, **dask**, and **Jupyter**. Additionally, cloud computing resources are accessible through their home page.

Earth Observation Using Python: A Practical Programming Guide, Special Publications 75, First Edition. Rebekah B. Esmaili.

DOI: 10.1002/9781119606925.appendixC

- **PyTroll** is a community of developers who curate a collection of packages that prioritizes ease of use and modularization for processing Earth satellite data. In particular, the **pyresample** package (Section 8.3.2) are maintained this community. They also have a series of online tutorials with examples on using their package: https://satpy.readthedocs.io/en/latest/examples/index.html

NASA/GES DISC has released a set of Jupyter notebooks that provide short and clear instruction on ordering, displaying, and analyzing satellite data. These notebooks can be found on https://disc.gsfc.nasa.gov/ by searching for Jupyter Notebook in "How-To's." Some useful notebooks include:

- How to download a spatial and variable subset of Level 1B data using OPeNDAP
- How to read and plot NetCDF MERRA-2 data in Python
- How to Read IMERG Data Using Python
- How to calculate and plot wind speed using MERRA-2 wind component data using Python
- How to Use the Web Services API for Subsetting
- How to Use the Web Services API for Subsetting MERRA-2 Data
- How to Use the Web Services API for Dataset Searching
- How to calculate greenhouse gas growth rates

Appendix D

TOOLS

D.1. Text Editors and IDEs

A few suggested IDEs are listed below:

- **PyCharm.** An IDE that provides code completion, error highlighting, and integration with iPython Notebook. This is a great Python-centric IDE.
- **Sublime Text.** A free, lightweight, and multiplatform text editor. Sublime has many add-on packages as well and one of the most popular ones online.
- **Atom.** Multiplatform and has a large number of add-on packages to enhance your code environment. Atom is owned by Github, so it has good integration with version control. Caveat: it uses a lot of memory.
- **Notepad++.** Windows only. It is simpler to use than Atom and does not use as much memory, as it uses less CPU power (which is only a concern on much older computers).
- **XCode.** MacOS only. This is a fully loaded IDE, so it contains both a text editor and a command line to edit and run code. This is a good choice if you want to create applications in addition to performing scientific analysis.
- **Visual Studio Code.** Multiplatform with extension support and an integrated terminal. It has a built-in command line so you can run the code within the editor and delivers overall good performance. Like Atom, it uses a lot of memory and has a slow launch time.

D.2. Terminals

To run shell scripts, you will need to open a terminal. A few terminal options are listed below:

- **CMD Prompt.** The base terminal for Windows.
- **PuTTy.** The most popular terminal emulator for Windows. Setup can be difficult for beginners, but it is very useful for connecting to remote servers.

Earth Observation Using Python: A Practical Programming Guide, Special Publications 75, First Edition. Rebekah B. Esmaili.

DOI: 10.1002/9781119606925.appendixD

- **Terminal.app.** This is native to the MacOS operating system. While initially Terminal was basic in older versions of MacOS, it has incorporated many of the features of popular third-party applications. A benefit is that it is already installed on MacOS and usually uses fewer resources than third party applications.
- **iTerm2.** A third-party terminal emulator for MacOS, with features such as autocomplete and multiple command lines, either by splitting the window or by opening tabs.

Appendix E

FINDING, ACCESSING, AND DOWNLOADING SATELLITE DATASETS

To obtain satellite data, it is common for most agencies to have a data-ordering process. Similar to an online merchant, you shop for your data by searching and adding datasets to your cart. There is no single repository for all data and must require you to register. However, as these services grow more sophisticated, you can subset your data and only download files you need rather than large (and unmanageable) quantities of data.

For near real-time (NRT) data, there are also FTP services to get access sooner and do not require the manual selection of data. These services include the Precipitation Processing System (PPS) https://pps.gsfc.nasa.gov/. However, not all data are rapidly available, and all have some form of data **latency**, which is the time between when the observation is made to when data are received by the user.

Below I will show the manual, checkout process for EarthData and CLASS, which are NASA and NOAA's primary data portals at the time this book was written. Like all web resources, they may evolve over time. All the HDF and netCDF data used in this book are real and accessible through the resources below. As an exercise, pick a file and try to find and order it online.

E.1. Ordering Data from NASA EarthData

- Go to https://earthdata.nasa.gov/
- Either browse for data or search. I will search for IMERG data as an example. Type "IMERG" into the search box and click on the search icon (🔍) to the right.

Earth Observation Using Python: A Practical Programming Guide, Special Publications 75, First Edition. Rebekah B. Esmaili.

DOI: 10.1002/9781119606925.appendixE

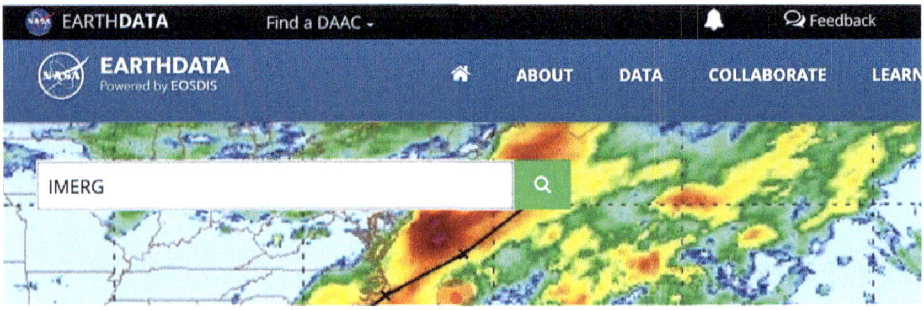

- Select GPM IMERG Final V00x, where x is the highest number available. The V stands for the version number, so selecting the highest number will be the newest version.

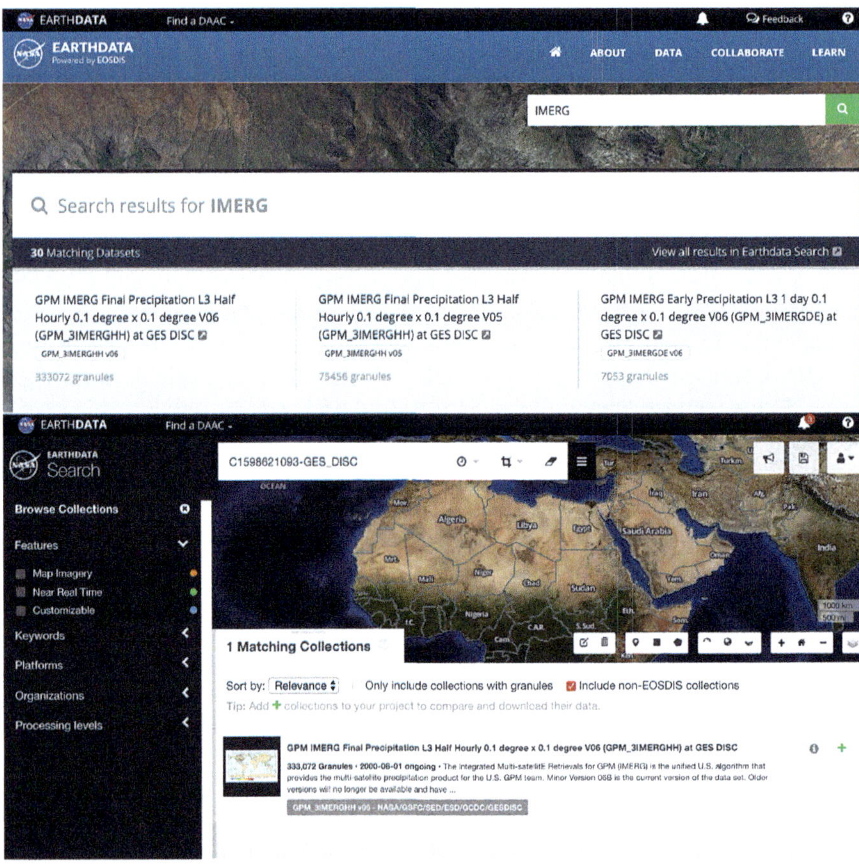

- You can subset by time and location using the tools in the top of the menu. For example, below I have made a bound around the Gulf of Mexico and selected dates from August 25–28, 2017, when Hurricane Harvey made landfall in Texas.

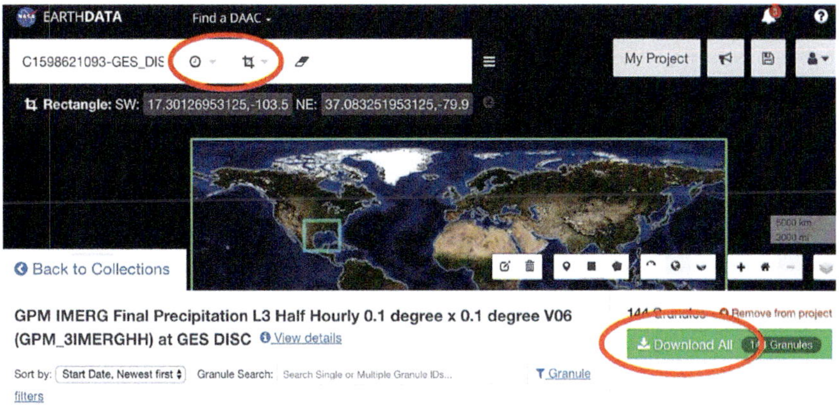

- When you are done, click the `Download All` button.
- You will be prompted to enter your username and password. If you have not yet registered, you can do this now.

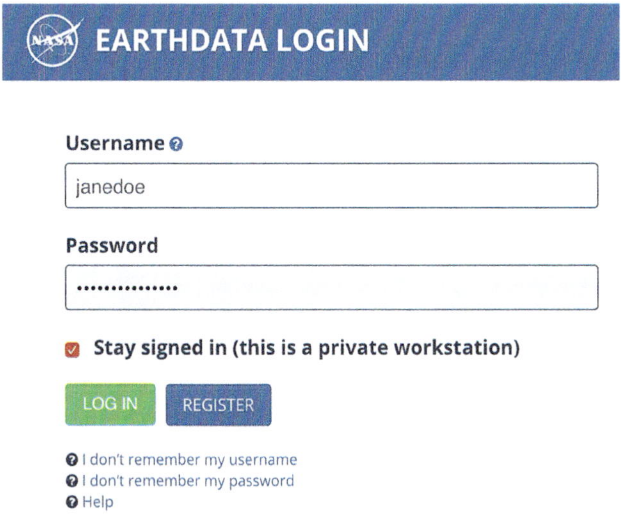

- You will be shown all available download options. Click done if you are ready.

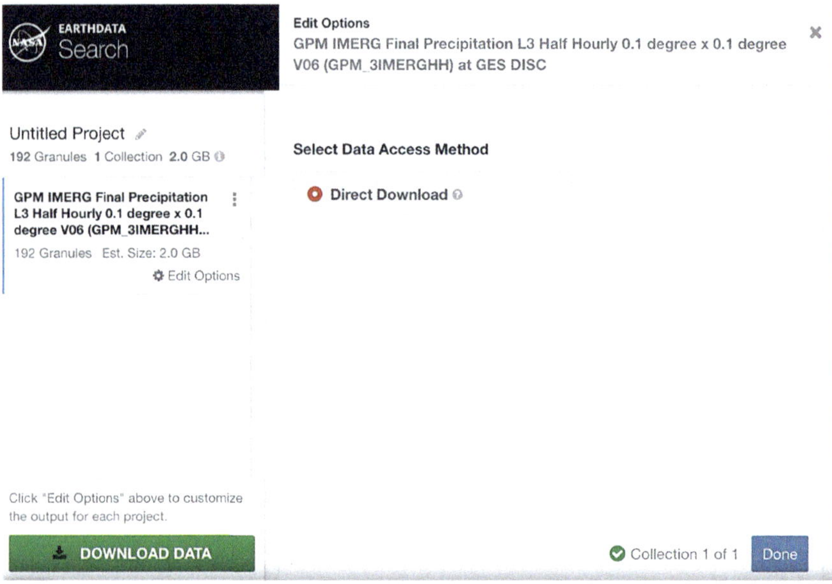

- The order has been placed, and you have been given an ID number. Data are not always immediately available for download, which can be between a few minutes to hours, depending on the size of the order.

E.2. Ordering Data from NOAA/CLASS

- Go to https://www.class.noaa.gov
- Click on the drop-down menu (many options will show up).

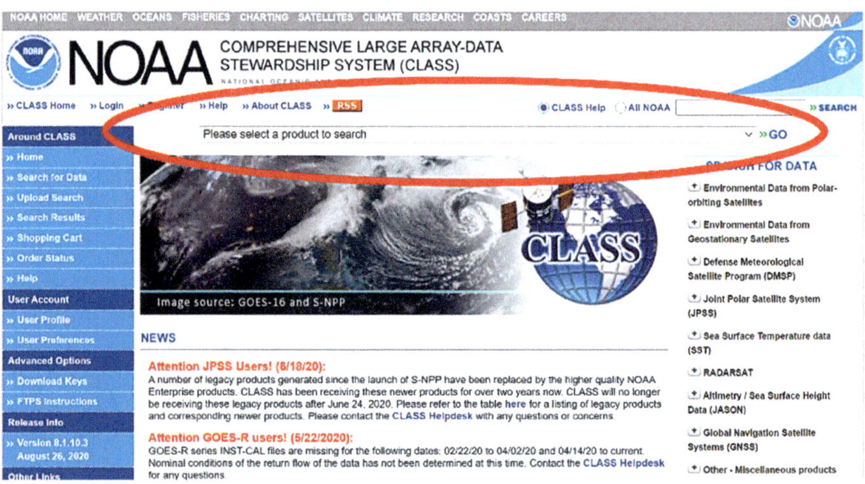

- Select JPSS Sounder Products (`JPSS_SND`) from the dropdown menu.

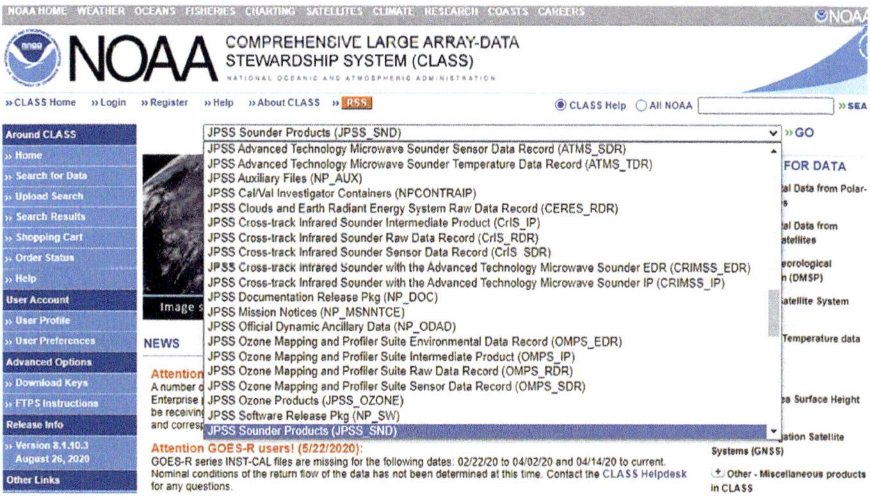

- Click `Go`

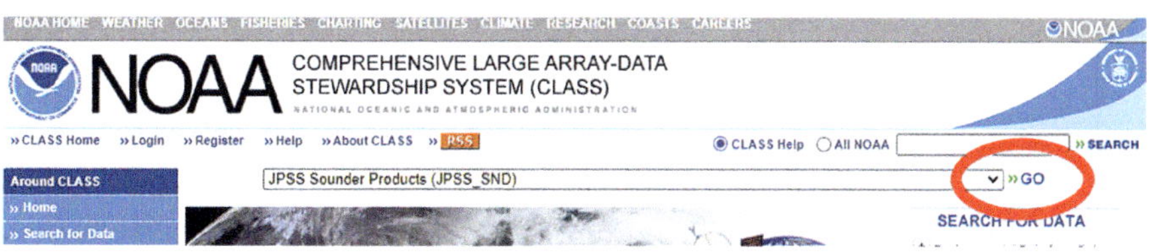

- Enter your search parameters, e.g., dates, locations. As an example, I selected the latitude and longitude bounds to be 15–35°N and 100–75°W. Under `Datatype`, I selected NUCAPS EDR. For the `Satellite` field, I selected Suomi NPP. Note that NUCAPS data are available from Suomi NPP from 2012 onward and from NOAA-20 beginning in March 7, 2019.

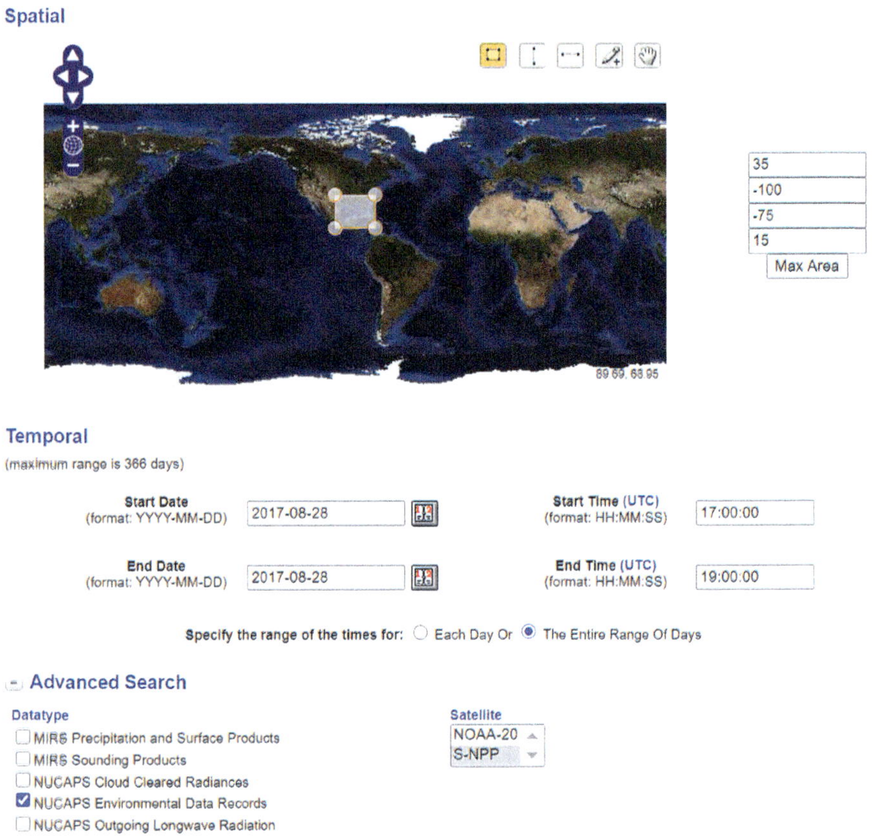

- Below you have two ordering options. `Search` allows you to review before ordering. `Quick Search & Order` skips to the last step without prompting you to review the search results.

- If you use `Search`, you will see a list of results. Click on the `Select Datasets` dropdown menu, click `1-13`, then click `Goto Cart`. Note: You can only order up to 500 datasets per order and will need to make a separate order if there are more or use the search option.

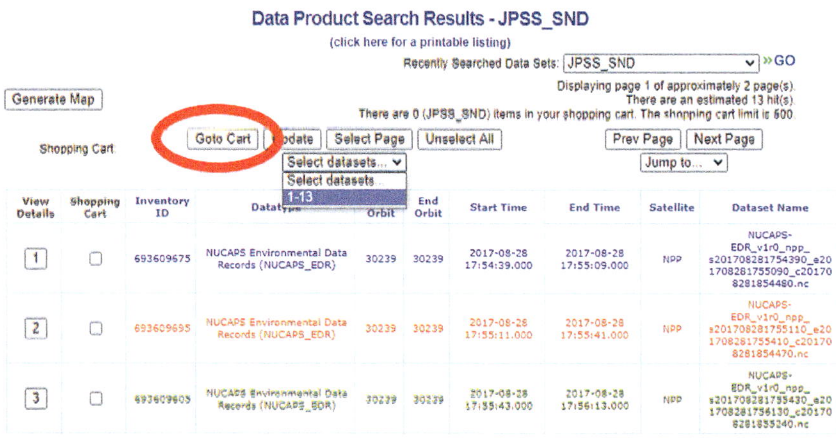

- If it is your first time ordering from CLASS, you will need to click `Register` and enter your name and email and create a password. If already registered, click `Login` at the top.

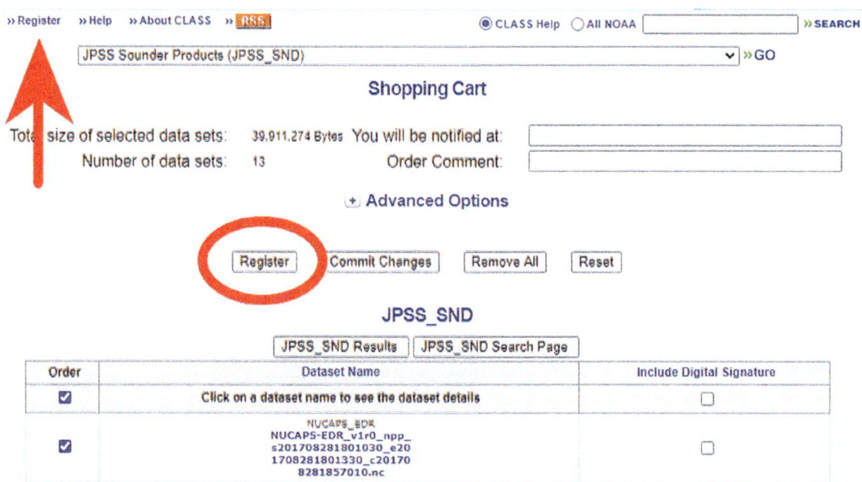

- Once you are logged in, the `Register` button will say `Place Order`. Click on place order.

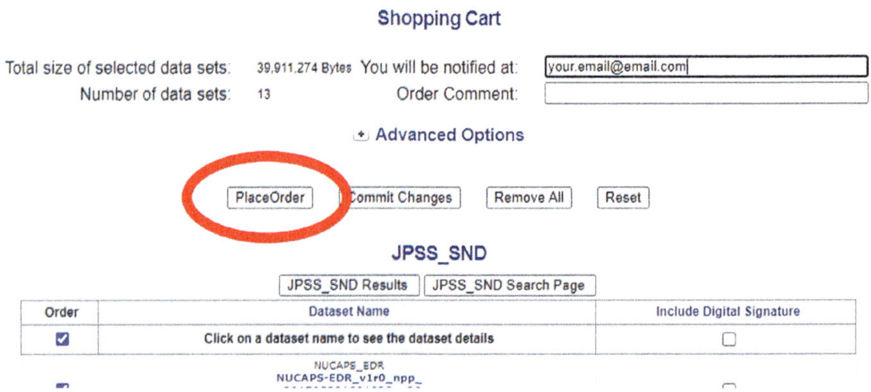

- After placing your order, you will receive an order confirmation email. Once the data have been processed, you will receive a second email notifying you that the data are available for download. This email will also include details on how to download the data via ftp or direct link.

Appendix F

ACRONYMS

Acronym	Definition
ABI	Advanced Baseline Imager
AHI	Advanced Himawari Imager
AOD	aerosol optical depth
API	Application Programming Interface
ARD	analysis-ready data
ASCII	American Standard Code for Information Interchange
ATBD	Algorithm Theoretical Basis Documents
ATMS	Advanced Technology Microwave Sounder
BCM	binary cloud mask
BUFR	Binary Universal Form for the Representation
CF	climate and forecast (Metadata)
CLASS	Comprehensive Large Array-data Stewardship System
CMD	command
CNES	Centre national d'études spatiales (National Centre for Space Studies)
COSMIC	Constellation Observing System for Meteorology Ionosphere and Climate
CPU	central processing unit
CrIS	Cross-track Infrared Sounder
CSPP	Community Satellite Processing Package
DAAC	Distributed Active Archive Center
DB	direct broadcast
DSR	Downward Shortwave Radiation
ECMWF	European Centre for Medium-Range Weather Forecasts
EOSDIS	Earth Observing System Data and Information System
ERSST	Extended Reconstructed Sea Surface Temperature
ESA	European Space Agency
ESRL	Earth System Research Laboratories
EUMETSAT	European Organization for the Exploitation of Meteorological Satellites
EXIS	Extreme Ultraviolet and X-ray Irradiance Sensors
FOR	field of regard

Earth Observation Using Python: A Practical Programming Guide, Special Publications 75, First Edition. Rebekah B. Esmaili.

DOI: 10.1002/9781119606925.appendixF

FOV	field of view
FOSS	free and open-source software
FTP	File Transfer Protocol
GCM	Global Climate Model or Global Circulation Model
GEO	geostationary
GFS	Global Forecast System
GIF	Graphics Interchange Format
GIS	Geographic Information Systems
GLM	Global Lightning Mapper
GMI	GPM Microwave Imager
GNU	GNUs Not Unix
GOES	Geostationary Operational Environmental Satellite
GPL	general public license
GPM	Global Precipitation Mission
GRACE	Gravity Recovery and Climate Experiment
GRACE-FO	Gravity Recovery and Climate Experiment Follow On
GRIB	GRIdded binary
GUI	graphical user interface
HDF	hierarchical data format
HRRR	High-Resolution Rapid Refresh
ICESat	Ice, Cloud, and land Elevation Satellite
ICOADS	International Comprehensive Ocean-Atmosphere Data Set
IDE	Integrated Development Environment
IDL	Interactive Data Language
IEEE	Institute of Electrical and Electronics Engineers
IMERG	Integrated Multi-satellitE Retrievals for GPM
IMSAI	Information Management Sciences Associates Inc.
IP	internet protocol
IR	infrared
JAXA	Japan Aerospace Exploration Agency
JPSS	Joint Polar Satellite System
JRR	JPSS Risk Reduction
LEO	low-Earth orbit
MATLAB	matrix laboratory
MERIS	MEdium Resolution Imaging Spectrometer
METAR	METeorological Aerodrome Reports
MIT	Massachusetts Institute of Technology
MODIS	Moderate Resolution Imaging Spectroradiometer
MOPITT	Measurement of Pollution in the Troposphere
MSG	Meteosat Second Generation
NAM	North American Mesoscale Forecast System
NASA	National Aeronautics and Space Administration
NDVI	normalized difference vegetation index
NESDIS	National Environmental Satellite, Data, and Information Service
NetCDF	Network Common Data Form
NOAA	National Oceanic and Atmospheric Administration

NRT	near-real time
NUCAPS	NOAA Unique Combined Algorithm Processing System
NWS	National Weather Service
OLCI	Ocean and Land Color Instrument
oPeNDAP	open-source project for a Network Data Access Protocol
PC	personal computer
PDF	portable document format
PEP	Python enhancement proposals
PNG	portable network graphics
PPS	Precipitation Processing System
RAM	random access memory
RGB	red green blue
SEVIRI	Spinning Enhanced Visible and Infrared Imager
SLSTR	Sea and Land Surface Temperature Radiometer
SMAP	Soil Moisture Active Passive
SST	sea surface temperature
STAR	Center for Satellite Applications and Research
SURFRAD	Surface Radiation Budget Network
SUVI	Solar Ultraviolet Imager
SVG	scalable vector graphics
UCAR	University Corporation for Atmospheric Research
UI	user interface
URL	uniform resource locator
UTC	Coordinated Universal Time
UTF	Unicode Transformation Format
VIIRS	Visible Infrared Imaging Radiometer Suite
VPN	virtual private network
WGS	World Geodetic System
WMO	World Meteorological Organization
YAML	YAML Ain't Markup Language

INDEX

Earth Observation Using Python: A Practical Programming Guide, Special Publications 75, First Edition. Rebekah B. Esmaili.

DOI: 10.1002/9781119606925.index